AF814343

Reproductive Biotechnology in Buffalo

Reproductive Biotechnology in
Buffalo

PREM SINGH YADAV, Ph.D.
Principal Scientist
Buffalo Physiology and Reproduction Division
Central Institute for Research on Buffaloes, Hisar-125 001

BIRBAL SINGH, Ph.D.
Senior Scientist
Animal Biotechnology Group
Indian Veterinary Research Institute
Regional Station, Palampur-176 061

INDERJEET SINGH, Ph.D.
Principal Scientist & Head
Buffalo Physiology and Reproduction Division
Central Institute for Research on Buffaloes, Hisar-125 001

R.K. SETHI, Ph.D.
Director
Central Institute for Research on Buffaloes
Hisar-125 001. INDIA

2010

SSPH

SATISH SERIAL PUBLISHING HOUSE
403, Express Tower, Commercial Complex, Azadpur, Delhi-110033 (India)
Phone : 011-27672852, Fax : 91-11-27672046
E-mail :info@satishserial.com, hkjain1975@yahoo.com
Website : www.satishserial.com

Published by :

SATISH SERIAL PUBLISHING HOUSE

403, Express Tower, Commercial Complex, Azadpur, Delhi-110033 (INDIA)
Phone : 011-27672852 Fax : 91-11-27672046
E-mail : info@satishserial.com, hkjain1975@yahoo.com

in association with

CENTRAL INSTITUTE FOR RESEARCH ON BUFFALOES
(Indian Council of Agricultural Research)

Hisar

© Director, CIRB, Hisar

ISBN 81-89304-80-1

© 2010. All rights reserved, no part of this publication may be reproduced, stored in a retrieval system or transmitted in any form or by any means, electronic, mechanical, photocopying, recording or otherwise without the prior written permission of the Director, CIRB and Publisher.

This book contains information obtained from authentic and highly regarded sources. Reasonable efforts have been made to publish reliable data and information, but the authors/editor(s)/contributors and publisher cannot assume responsibility for the validity of all materials or the consequences of their use. The authors/editor(s)/contributors and publisher have attempted to trace and acknowledge the copyright holders of all material reproduced in this publication and apologize to copyright holders if permission and acknowledgements to publish in this form have not been given. If any copyright material has not been acknowledged please write and let us know so we may rectify it.

Typeset at: **Laxmi Art Creations** # 98 11 48 23 28
Printed at: Salasar Imaging Systems, Delhi-35

Preface

From Darwin to Dolly, domesticated livestock have guided our understanding of evolution and biology. The domestic buffalo (*Bubalus bubalis*) is a multipurpose livestock species in many countries of the world, particularly in South Asia, Mediterranean region of Europe and South America, and is an indispensable source of employment to the marginal farmers and landless labourers. India has more than half of the world buffalo population and represents a few superior genotypes in the form of Murrah, Nili Ravi and Jaffarabadi breeds apart from a number of non-descriptive animals adapted to the prevailing agroclimatic niches. The buffaloes exhibit high feed conversion efficiency and disease resistance compared to cattle and other livestock species. High-yielding Indian buffalo breeds, such as Murrah, can yield as high as 4000 L of milk over 300 days lactation. There is still a room for enhancing milk production through selective breeding, augmented rumen fermentation, improved management and disseminating the elite genotypes. High protein and fat contents in milk will keep the buffaloes at forefront of dairy applications and meat production besides their use for skin and draught power.

Poor reproductive performance is an emerging trend in dairy farms. Weak and silent estrus and embryonic or fetal losses are among some of the major causes of reduced buffalo production.

Cutting-edge reproductive technologies promise numerous possibilities for conservation and production of valuable buffalo breeds. Embryo biotechnology offers enormous potential by using low-yielding (but reproductively efficient) females as surrogate mothers for the production of elite calves through embryo transfer. During the past decade remarkable achievements have been made in the field of buffalo assisted reproduction. China and India have been successful in producing the cloned buffalo calves. It is high time now to improve the technology for wider application under field conditions to disseminate the valuable germplasm and conserve the endangered indigenous buffaloes capable of breeding and producing milk under low input conditions.

This book contains review of commonly employed laboratory procedures and includes perspectives on how to maximize success rates of less efficient technologies.

Authors

Acknowledgements

It is our pleasure to thank all those who helped directly or indirectly in preparation of this book. Writing book of this nature would not have been possible without consulting research papers and reviews of many researchers and authors in this field, authors are indebted to all of them.

We are highly thankful to Dr. K.P. Agrawal, former National Coordinator, NATP/ NAIP (World Bank Aided Projects) ICAR, New Delhi and Dr. R.K. Singh, Director, National Research Center on Equines, Hisar for their editorial comments and guidance to strengthen the manuscript technically on both the aspects of reproduction biotechnology and generating readers' interest in this upcoming topic for buffalo improvement.

We are also thankful to Dr. O.P. Sharma, Officer Incharge, Indian Veterinary Research Institute Regional Station, Palampur for his kind support and valuable suggestions.

We are also thankful to our friends, colleagues, members of Publication Committee at CIRB, Hisar for their unstinted support and suggestions for the publication of this book. Last but not the least we are thankful to the publishers for making this book a reality.

Authors

Contents

1

Water Buffalo-An Asset Unraveled

"Indian subcontinent is the home tract of world elite buffaloes. The major research areas aimed to augment buffalo production include reproduction, feed utilization and disease management. Current needs are proliferation of elite genotypes and endangered indigenous breeds"

Introduction

The water buffalo (*Bubalus bubalis*) is a semi aquatic animal of the bovine family. There are about 158 million buffaloes in the world, roughly 97% of which are water buffaloes essentially found in the Asian Region. The buffaloes are predominantly kept for milk production, but also used for meat, fuel and fertilizer (from manure), as well as draught power **(Table-1)**. The domesticated water buffalo is often referred to as the "living tractor of the East," as it is relied upon for ploughing and transportation in many parts of Asia. Demand for buffalo milk and meat has expanded due to growth of the human population, and overall increase in income level (www.aphca.org/publications/files/w_buffalo.pdf...23.10.2009).

There are at least nine well-defined buffalo breeds in India, representing approximately 30% of the total 90 million buffalo population, the remaining animals are non-descript (George *et al.*, 1988). Importance of buffalo to dairy industry is immense as they constitute 35% of the bovine population in India and contribute more than 55% to the total milk production in the country. In addition, each year, an adult buffalo produces 4 to 6 tons of wet manure plus additional urine as bio-fertilizer to the land. This may minimize the need for chemical fertilizers and may provide essential soil humus. The buffalo rumen microbiome appears to possess higher capabilities of degrading the fiber-enriched forages and supplying the host optimal nutrition. Rumen microbes from buffaloes may have important industrial applications as well.

General classification and distribution

Buffaloes have descended from *arni* or wild Indian buffalo. The world has two main types of buffaloes viz., i) Riverine and ii) Swamp, whereas the third, Mediterranean type buffaloes have emerged from these major genotypes.

The river buffaloes

They are usually large size animals with black skin, long curled horns and hold their heads high **(Fig. 1)**. The average body weight of adult river buffalo is about 800-1000 kg in mature male and 700-900 kg in mature female. They are primarily of dairy type animals reared for milk, but also used for meat production. The river buffaloes are distributed in South Asia and South West Asia. Approximately 70% of the total world population of river buffalo is reared in high numbers in South Asia. India is the leading buffalo rearing country producing 96 million tons of buffalo milk (with 5% annual growth) and 1.58 million tons of buffalo meat. The demand of buffalo milk for dairy products and local consumption is increasing. Within the Riverine type, there are several breeds (e.g. Murrah, Nili-Ravi, Surti, etc.) that have been developed over the centuries to meet human's needs in different environments. However, the swamp buffaloes have not been differentiated, and the entire population represents one and the same breed, even though the animals are geographically isolated.

The swamp buffaloes

The swamp buffalo is of small size with compact body and has straight horns, and primarily used as draught animals **(Fig. 2)**. The average body weight of swamp buffalo is about 400-600 kg in mature male and 350-450 kg in mature female. The swamp buffaloes are distributed in Laos, Cambodia, Vietnam, Taiwan, China, Philippines and Indonesia. They are usually dark grey but may also be black, black and white, or even all white, and have long, gently curved horns. Swamp buffaloes have a chevron on the chest, legs lighter in colour and larger hooves.

Swamp buffalo is a late maturing animal reaching sexual maturity after 3 years of age under average conditions of nutrition and management. The female reproductive life extends from 3-16 to 25 years of age. The gestation period is around 325-330 days, and milk yield is low being 1-2 liters per day. The swamp buffalo is considered an excellent draft animal because of big feet and excellent pulling capacity and slow walking steps. Riverine and swamp buffaloes can be crossbred. However, some of their offspring may be infertile because of difference in chromosome numbers of river and swamp buffaloes.

Mediterranean buffaloes

The third type i.e. Mediterranean buffaloes **(Fig. 3)** are abundant in Italy, the Balkan States, Romania, Brazil, Turkey, a few hundred in the United Kingdom, Germany, Netherland, Switzerland, Hungary and in some parts of Russia. The Mediterranean buffalo originated from the Indian buffalo. It was introduced into Europe with the advent of Islam and the Arab occupation as well as through other central European conquerors in the 6th and 7th Centuries.

Fig. 1. Murrah, water buffalo

Fig. 2. Swamp buffalo

Fig. 3. Mediterranean buffalo

They are generally black, black and brown or dark grey in colour. Horns are flat at the bottom, turned backwards and slightly outwards with the tip pointed inwards. They have a compact conformation with a deep and wide chest as well as a developed pectoral. The back and rump are short. The udder is medium in size with squarely placed quarters and halves; the teats are cylindrical. Size, body weight and productivity vary a lot according to the environment and management. The average body weight of the adult female is 450-650 kg. Average daily milk yield reveals a huge variability, mainly depending on the feeding system. It can range from 3 to 4 kg milk/day for poorly fed animals to 15 kg/day in intensive management systems (ftp://ftp.fao.org/docrep/fao/010/ah847e/ah847e01.pdf., accessed Nov. 5, 2009)

Domestication of water buffalo

Water buffaloes have even found a mention in Hindu lore, as the god of death "Yama", rides on a water buffalo (Bakkannavar *et al.*, 2009). The present day river buffalo was domesticated independent of swamp buffalo (Kumar *et al.*, 2007a) and are the result of complex domestication processes involving more than one maternal lineages and a significant maternal gene flow from the wild populations following their initial domestication events (Kumar *et al.*, 2007b). Molecular and phylogeography data are consistent with the available archaeological information in supporting the proposition that the river buffalo was likely to be domesticated in the Western region of the Indian subcontinent, specially the present tracts of the Mehsana, Surti and Pandharpuri breeds (Kumar *et al.*, 2007b). It is concluded that the Mediterranean and the Indian domestic river buffaloes have been derived from the same stock through a complex process, and that during domestication and/ or afterwards there might have been a continuous influx of maternal variability from the wild buffalo into the domestic stocks (Kumar *et al.*, 2007b).

The water buffalo and its multipurpose use

As food animal

India has largest livestock population in the world. It has about 98 million buffaloes, which is 57% of total population in the world, contributing to 1.48 million metric tons of meat (FAO, 2008). The increase in meat and milk production is being demand-driven although is not evenly spread among developing countries. The buffalo meat is the healthiest meat among red meats known for human consumption due to low calories and cholesterol (Kandeepan *et al.*, 2009a), and is comparable to beef in many of its physicochemical, nutritional and functional properties and sensory attributes. Furthermore, its use in meat processing is increasing because of its higher content of lean meat and lower fat (Kandeepan *et al.*, 2009b). The buffalo meat has gained importance in the recent years because of its domestic needs and export potential. Buffaloes properly managed and raised as meat animal and slaughtered at 16-20 months of age yield highly satisfactory top quality meat at much lower cost than the cattle (Ranjan and Pathak, 1979). Moreover, the demand for meat from other species such as pigs and poultry could be met by intensive production. Unlike other meat producing animals (cattle and pig), the buffalo has no religious taboos, and is slaughtered all over the country. Buffalo meat should not merely be regarded as a cheap source for meat products, but should be valued as high quality meat, especially if derived from young animals.

As an important draft animal

Water buffalo is an integral part of traditional village farming structure and due to their natural adaptation to low-input conditions **(Table-1)** and remarkable feed conversion efficiency. The domesticated water buffalo is often referred to as "The living tractor of the East" as it is mainly relied upon for ploughing and transportation in many parts of Asia. Probably the most adaptable and versatile of all work animals, the buffalo is widely used to plough; level land; plant the crops; puddle rice fields; cultivate field crops; pump water; haul carts; sleds; and shallow-draft boats; carry people; thresh grain; press sugar cane; haul logs; and much more (Bakkannavar *et al.*, 2009). The buffaloes compared to cattle and other livestock species are more useful as draft animals.

Threats to buffalo genetic biodiversity

Economic and market-driven threats, inadequate livestock-sector policies, poor conservation strategies, inadequate institutional capacities to manage breeds and loss of labour, are the major threats eroding livestock genetic diversity. Threats to livestock diversity and decline in their populations (Table-2) include decreases in public funding, lack of political will to support rural communities, inappropriate policies and legislation including environmental schemes, disease, predators, urbanization, poor return on product, competition from other livestock, ageing of the farming population, lack of marketing support, inbreeding in animal populations and loss of skills (Carson *et al.*, 2008). Livestock keepers, specifically those who depend on natural forage resources to raise their animals, have to cope with many challenges. This makes the livestock rearing more difficult. Like other domestic animals, some of the buffalo breeds are threatened and may become extinct in future.

Opportunities and challenges

The world's domestic animal diversity comprises some 7,616 documented mammalian and avian breeds as well as an unknown number of not yet documented breeds. These breeds are largely the result of livestock breeding strategies managed by farmers and pastoralists. Global output of the livestock sector particularly in developing countries is expected to double in the next 25 years. It is essential that most of this increase is achieved through improved efficiency rather than expanded numbers. A range of biotechnologies have important contributions to achieve these objectives. While crops have a variety of reproductive mechanisms which can facilitate validating new technologies, animals have only one reproductive mechanism. This makes access to and intervention with the genetic material difficult and expensive. Much animal biotechnology is, therefore, centered on managing the reproductive process, and particularly on aspects of embryo production. However, commercial applications of these technologies are unlikely to be available for several years, until the cost is reduced. Nevertheless, direct identification of genes affecting important traits for economic performance or disease resistance together with the use of genetic markers could accelerate genetic improvement.

Conclusion

The innate potential of buffaloes to meet present and future nutrient needs of rapidly growing population deserves special attention. Reproductive biotechnology validated in other farm animals can be exploited in various ways (Table-3) to augment bubaline production. Private entrepreneurs are required to be technically and financially supported for production of quality forage at larger scale to overcome the forage scarcity. The reproductive technologies need to be refined for application under field conditions.

Table 1. A summary of benefits of raising water buffaloes as livestock species

1. Ability to subsist on a low quality, high roughage diet, and excellent body weight gain compared to other species.
2. Resistance to most internal and external parasites and diseases that affect cattle (Ross, 1975).
3. Generally low birth weight and low incidences of calving problems.
4. Quick and easy calf growth due to high percentage of butterfat and milk solids in the dam's milk.
5. Growth rate comparable to cattle. They are reasonably productive up to 15-18 years of age (Banerjee, 1998).
6. Meat flavor comparable to beef with about one-half the cholesterol, and less than one-fourth the amount of fat; no religious hindrances in slaughtering the buffaloes
7. Milk, with an average of 8 % butterfat, is used in the production of mozzarella cheese, other dairy products and ethnic sweets.
8. Requirements for fencing, handling equipment, inoculations, and health programs are similar to those of cattle.
9. Offers good draught power for tropical and humid environment
10. Dung as fuel and manure for improving soil fertility.

Table 2. General factors responsible for erosion of livestock biodiversity

Factors	Description
Anthropological developments	Preferences given to high-input, high out-put breeds developed for benign environments; economic interests of commercial companies to disseminate their products irrespective of the genetic capability of the breeds to adapt in new environments; shifting to other professions for short-term benefits and giving up livestock domestication; replacements of animals by applications of machinery in crop harvesting, ploughing and transportation etc.
Biased selection	Emphasis being given on a single trait, e.g. milk production or meat production, which may lead to rejection of traditional or multipurpose animals by the farmers and livestock owners
Genetic introgression	Indiscriminate crossbreeding strategies and accidental introgression leading to loss of pure indigenous livestock breeds
Biotechnological interventions	Applications of advanced ARTs like nuclear transfer cloning, and transgenesis imposing additional pressure on the physiology and well-being of the animals; inefficiency of certain technologies like cryopreservation to store and disseminate the valuable germplasm
Environmental factors	Abrupt changes in climatic conditions, natural disasters like flood, draught and global warming adversely affect the livestock populations and their production efficiencies, loss of vegetation or emergence of new weeds and phytotoxins in the dietary resources of animals
Lack of proper breeding policies	Replacement or dilution of indigenous pure breeds by cross breeds, failure of crossbreeds to adapt to prevailing low-input conditions, higher susceptibility of exotic breeds parasitic infestations etc.

Table 3. Proposed strategies to improve buffalo production

1.	Improving access of marginal farmers and livestock owners to indigenous buffalo breeds
2.	Collecting and effectively disseminating more and authentic information on farming systems (particularly with regard to economical and sustainable production) to farmers and other concerned parties
3.	Further developing "model buffalo rearing villages" based on best-practices
4.	Providing incentives for production of quality meat and meat products from buffaloes
5.	Developing breeding schemes to improve buffaloes for long-term and sustainable use
6.	Developing appropriate biotechnology innovations specific to buffaloes. Possible areas are artificial insemination, embryo production and transfer under field conditions, sperm sexing and its larger scale applications, animal cloning, and use of biotechnology to improve feed utilization and disease control.

References

Bakkannavar SM, Monteiro FNP, Bhagavath P and Kumar GP. **2009**. Death by attack from a domestic buffalo. Case Report. *Forensic Legal Medicine.* (in press).

Banerjee GC. **1998**. Buffaloes. **In**: *A Textbook of Animal Husbandry*. 8th edition, pp. 699-700, 711-712.

Carson A, Elliot M, Groom J, Winter A and Bowles D. **2008**. Geographical isolation of native sheep breeds in the UK - evidence of endemism as a risk factor to genetic resources. *Livestock Sci.* **123**, 288-299.

FAO. 2008. The state of food and agriculture. Food and Agriculture Organization, Rome. (http://apps.fao.org.page/cpllection).

George M, Balaine DS, Vij PK and Kumar S. **1988**. Conservation and management of buffalo genetic resources of India. **In**: *Buffalo Production and Health*. R. Nagarsankar (Ed.). ICAR, New Delhi, India, pp 31-37.

Kandeepan G, Biswas S and Rajkumar RS. **2009a**. Buffalo as a potential food animal. *Int. J. Livestock Prod. Sci.* **1**. 1-5.

Kandeepan G, Anjaneyulu ASR, Kondaiah N, Mendiratta SK and Lakshmanan V. **2009b**. Effect of age and gender on the processing characteristics of buffalo meat. *Meat Sci.* **83**, 10-14.

Kumar S, Nagarajan M, Sandhu JS, Kumar N, Behl V and Nishant G. **2007a**. Mitochondrial DNA analyses of Indian water buffalo support a distinct genetic origin of river and swamp buffalo. *Anim. Genet.* **38**, 227-232.

Kumar S, Nagarajan M, Sandhu JS, Kumar N and Behl V. **2007b**. Phylogeography and domestication of Indian river buffalo. *BMC Evol. Biol.* **7**, 186.

Ranjan SK and Pathak NN. **1979**. Management and Feeding of Buffaloes. Vikas Publishing House, New Delhi (India), pp 217-227.

Ross C. **1975**. Alternative livestock: with particular response to the water buffalo (*Bubalus bubalis*). **In**: *Meat*. 1st Publication, pp. 507-524.

—□□□—

2

Assisted Reproductive Technologies-A Historical Perspective

"The buffalo is probably the last livestock species to have been domesticated, with many genetic, physiological and behavioral traits not yet well understood. In view of its economic contribution to the rural livelihood and dairy and meat applications, and adaptation to low-input and extreme climatic condition, the water buffaloes have emerged as indispensable livestock species. During the recent years, the water buffalo has attracted a huge attention of the reproductive biologists and livestock researchers"

Introduction

Some 12000 to 14000 years ago, humans began to domesticate animals and plants and to spread their Neolithic farming across most of the world's terrestrial habitats. Thousands of years of natural and human selection, genetic drift, inbreeding and crossbreeding have led to the diversity of animal genetic resources we see today, which in turn have sustained animal agriculture in diverse environments and production systems. Remarkably few animal species were ever successfully domesticated. Only 14 of the 40 or so domesticated mammalian and bird species provide 90 per cent of human food supply. Other livestock species (water buffaloes, yaks, horses, asses, camels, llamas, reindeers, ducks, geese and turkeys) have smaller populations but are crucial to the livelihoods and survival of hundreds of millions of poor livestock keepers in developing countries and in marginal areas.

The domestication of livestock species- some ten thousand years ago, was a vital step in the development of human civilization through providing humans with their milk, meat, hides, strength, speed and companionship. Over the centuries, the domestication evolved into breeding and the genetic improvement of domesticated animals. Now a days, breeders measure many different animal traits and choose the best animals to be the parents of next generation. The way the livestock species are used for milk or meat production has changed during the last two decades. Now, livestock are supposed to give us even more, thanks to new insights from genomics and molecular phylogenetics.

Genetic improvement of livestock is a particularly effective technology, producing permanent and cumulative changes in performance. New biotechnologies are created and applied to the livestock industry to increase efficiency in both meat and dairy production system. These technologies are primarily optimized for augmenting the reproductive efficiency of elite females and disseminating the use of stud sires.

Assisted reproductive technologies (ARTs) in a livestock perspective can be defined as the techniques that manipulate reproduction-related events and/or structures to achieve pregnancy with the final goal of producing healthy offspring. In dairy production the ultimate aim of ARTs is the birth of healthy offspring and efficient utilization of females. In general, the ARTs manipulate the events/ structures related to ovulation, fertilization and embryo development. Though these techniques vary in their impact, the literature shows that ARTs are being currently applied not only in commercial livestock production settings, but also in rural settings. Whereas, some of these techniques are more widely applicable in disseminating the elite genotypes, the more advanced ones like somatic cell nuclear transfer (SCNT) and transgenesis have low efficiency in producing offspring and in some cases can cause serious detrimental effects at the prenatal and postnatal period. However, if improved and applied appropriately, they might be more useful in the conservation of endangered livestock species, larger-scale production of high quality foods and pharmaceuticals, and developing experimental models for human and biomedical applications.

Embryo transfer (ET) technology can be a valuable research tool, and can benefit a number of potential research topics that could benefit from having this technology. Such topics include livestock genetic selection, reproductive function, nutrition and disease transmission assessment. The technology is ideally suited to conserving indigenous buffalo population and maintaining the superiority in the prevailing socioeconomic and agro-climatic conditions. The technology provides enormous opportunities for livestock breeders to optimize genetic progression and increase the number of offspring borne from different bloodlines within a given time-frame.

Advances in ARTs - A historical overview

ET technology is a multi-factorial process that encompasses a series of integrated sequential steps including oocytes-donor selection, treatment of donors for superovulation, insemination of the superovulated donor, embryo recovery and evaluation, recipient selection, embryo transfer to the estrus-synchronized recipients, and the surrogate mother and neonate healthcare, etc.

The use of ET was first documented in 1880 by Walter Heape with the birth of two Angora rabbits from a British White doe following embryo transfer (Heape, 1891). Among the bovine species, the first ET cattle calf 'Virgil,' was produced around 60 years ago (Willet *et al.*, 1951). However, commercial ET programs started in North America during the early 1970s, primarily as a means of multiplying the numbers of calves produced by exotic breeds of beef cattle. However, the use of ET in dairy animals remained restricted owing to lack of technical skills for collection of oocytes or embryos. The advent of non-surgical collection of oocytes in mid-1970s, raised novel hopes for applications of technology at larger scales. Around the same time, the non-surgical embryo transfer techniques, based on the Cassou inseminating gun (Instruments de Medecine Veterinaire, IMV) were developed. Together, these advancements opened the ways to proliferate the ET for applications in cattle under field conditions. ET combined with the use of freeze-thaw protocols which permitted embryos to be shipped globally completed an important requirement for commercial exploitation of ET in cattle. Alberta Livestock was the first commercial company established in 1971that aimed at utilizing the ET at commercial scale for livestock production. Two years later, Wilmut and Rowson (1973) announced the birth of cattle calf (Frosty II) from the transfer of a frozen-thawed embryo. This was followed by the birth of cattle calf after transfer of embryos produced by IVF.

Originally, the multiple ovulation and embryo transfer (MOET) in combination with AI was initiated as a new breeding strategy in cattle (Nicholas, 1979). Thereafter, superovulation through exogenous hormonal administration was initiated among buffaloes for deriving genetically superior embryos with an increased rate. The process relied mostly on protocols that were originally applied for cattle (Singla and Madan, 1990). Superovulation followed by insemination of donor, embryo collection, and embryo transfer have allowed the researchers and dairy producers to obtain multiple offspring from only a few genetically superior (donor) females. At present the technology is used in a number of livestock species including porcine, horse, sheep and goat.

Embryo technologies in buffaloes were initiated in early 1960s, but actual outcome started in early 1980s. The success in ET program in buffalo began as early as in 1983 with first ever birth of a buffalo ET calf (Drost *et al.* 1983). Kamonpatana (1990) had reviewed the status of bubaline assisted reproduction at that time, and noted that ET in buffaloes, particularly the swamp buffaloes had been attempted in several countries, including Bulgaria, India, Malaysia, Pakistan, and USA. Subsequently, the technology was used to disseminate the buffaloes and birth of a live calf (called "Pratham") was reported in India from the transfer of IVF-derived embryos (Madan *et al.* 1991). Most attempts at producing buffalo embryos *in vitro* were based on the methods optimized and used in cattle (Singla and Madan, 1990; Gordon, 1994). Research on *in vitro* embryo production (IVEP) continued all over the world including India (Totey *et al.*, 1991; 1993a, b; Sharma *et al.*, 1994), Taiwan (Lu and Hsu 1991; Pavasuthipaisit *et al.*, 1992), Thailand (Chuangsoongneon and Kamonpatana 1991) and Malaysia (Jainudeen *et al.*, 1993).

The landmark developments in the past half-century leading to gametes biotechnology as being practiced currently have been covered in some excellent reviews (First *et al.*, 1999; Foote *et al.*, 1999a, b; Pressice, 2007; Singh *et al.*, 2009).

Table 1. Milestone events in livestock embryo transfer and related technologies and their applications in buffaloes

Year	Events (Reference)
1890	Birth of rabbits from ET (Heape, 1890)
1949	Birth of sheep lamb and goat kid from ET (Warwick and Berry, 1949).
1951	Birth of cattle from ET (Willet *et al.*, 1951)
1960	The two-wave concept for follicular growth during the bovine oestrus cycle was propounded (Rajakoski,1960).
1972	Development of slow-freezing methods of embryos, and births of live offspring from the cryopreserved embryos (Whittingham, 1971; Festing, 1972)
1979	Initiation of ET technology in cattle (Nicholson, 1979).
1981	Birth of "Virgil", the first calf produced by IVF in cattle (Brackett *et al.*, 1982)
1981	Establishment of technology for isolation of murine embryonic stem cells (Evans and Kaufman, 1981; Martin, 1981)
1982	Growth and de-differentiation of estrogen-active and estrogen-inactive follicle during the estrus cycle were distinguished (Ireland and Roche, 1982, 1983a, b)
1983	**Birth of world's first buffalo calf through embryo transfer (Drost *et al.*, 1983)**

1984 Establishment of ultrasound-based methods to monitor sizes of follicles during the oestrus cycle of heifers (Pierson and Ginther, 1984).

1986 Feasibility of Intergenic embryo transfer between water buffalo and cattle were explored, reciprocal transfer of buffalo embryos to water buffaloes were attempted (Drost *et al.*, 1986), induction of superovulation in buffalo ET program, and establishing non-surgical embryo-recovery procedures (16-20 gauge Foley catheter) in buffaloes (Drost *et al.*, 1986).

1987 Establishment of the concept of dominant follicles, as observed in primates, and it was applied to cattle, and the three-wave hypothesis for development of dominant follicles during the oestrus cycle was proposed. (Ireland and Roche, 1987).

1988 Embryo collection described in swamp buffaloes was described using either a two-way Foley catheter or an IMV embryo collector (Sharifuddin and Jainudeen, 1988)

1989 Cryopreservation of *in vitro* produced embryos (Techakumphu *et al.*, 1989)

1991 IVF calves born using the IVF technologies (Madan *et al.*, 1991; Suzuki *et al.*, 1992)

1993 Successful culmination of pregnancy and live birth from transfer of frozen-thawed embryos (Kasiraj *et al.* 1993), sex-determination of embryos by PCR amplification of DNA from blastomeres (Rao *et al.*, 1993)

1997 Production of "Dolly" (Wilmut *et al.*, 1997), *In vitro* embryo production from OPU-derived oocytes (Galli *et al.*, 1998), micromanipulation and cloning studies on oocytes and embryos (Singla *et al.*, 1997)

1999 Production of nuclear transfer cloned embryos (Parnpai *et al.*, 1999)

2001 Studies on effects of interspecies cytoplasmic recipients and activation protocols on nuclear transfer embryo production (Kitiyanant *et al.*, 2001)

2003 Pregnancies established from *in vitro* produced and vitrified embryos (Neglia *et al.*, 2003)

2004 Calves born from transfer of *in vitro*-derived and vitrified-thawed embryos (Hufana-Duran *et al.*, 2004)

2005 Pregnancies established after insemination with sexed sperm (Presicce *et al.*, 2005a), and calves born following AI with sexed semen (Presicce *et al.*, 2005b), vitrification of nuclear transfer cloned embryos (Laowtammathron *et al.*, 2005)

2006 **Rapid sexing of embryos using loop-mediated isothermal amplification (Hirayama *et al.*, 2006); initiation of the pioneering work on handmade cloning (HMC) in water buffaloes (Singh, 2006).**

2007 **First surviving cloned buffalo calves in buffalo (Shi *et al.*, 2007), birth of female twins from embryos produced through IVF using flow-cytometric sorted sperms (Lu *et al.*, 2007), first reports on ES cells from IVF-derived (Verma *et al.*, 2007), and parthenogenetically-produced embryos (Sritanaudomchai *et al.* 2007), and fetal embryogenic stem cells (Huang *et al.*, 2007)**

2008 **Birth of sex-preselected calves (Liang *et al.*, 2008); production of transgenic embryos expressing GFP (Verma *et al.*, 2008; Huang *et al.*, 2009; Wadhwa *et al.*, 2009)**

2009 **Studies on zona-free or HMC in water buffalo (Singh, 2006; Singh *et al.*, 2008; Shah *et al.*, 2008); establishing pregnancies from the zona-free embryos, (Shah *et al.*, 2009); cryotop vitrification of buffalo oocytes (Attanasio *et al.*, 2009); birth of water buffalo from transfer of HMC embryos (b2bportal.ohlog.com/garima-india.oh66736.html, accessed Nov. 4, 2009)**

(The achievements in buffalo ARTs are highlighted as bold)

References

Attanasio L, Boccia L, Vajta G, Kuwayama M, Campanile G, Zicarelli L, Neglia G and Gasparrini B. **2009**. Cryotop vitrification of buffalo (*Bubalus bubalis*) in vitro matured oocytes: effects of cryoprotectant concentrations and warming procedures. *Reprod. Domest. Anim.* (in press).

Brackett BG, Bousquet D, Boice ML, Donawick WJ, Evans FJ and Dressel MA. **1982**. Normal development following in vitro fertilization in the cow. *Biol. Reprod.* **27**, 147-158.

Chuangsoongneon, U and Kamonpatana M. **1991**. Oocyte maturation, *in vitro* fertilization and culture system for developing preimplantation swamp buffalo embryos using frozen thawed semen. *Buffalo J.* **7**, 189–198.

Drost M, Wright JM Jr, Cripe WS and Richter AR. **1983**. Embryo transfer in water buffalo (*Bubalus bubalis*). *Theriogenology.* **20**, 579–584.

Drost M, Wright JM and Elsden RP. **1986**. Intergenic embryo transfer between water buffalo and domestic cattle. *Theriogenology.* **25**, 13–23.

Evans MK and Kaufman MH. **1981**. Establishing in culture of pluripotent stem cells from mouse embryo. *Nature.* **292**, 154-156.

Festing MF. **1972**. Mouse strain identification. *Nature.* **238**, 351-352.

First NL, Mitalipova M and First MK. **1999**. Reproductive technologies and transgenics. **In**: *The Genetics of Cattle*. Fries R and Ruvinsky AJ (Eds). CAB International Wallingford, pp 411-436.

Foote RH. **1999a**. Artificial insemination to cloning. Tracing 50 Years of Research. Cornell University, Ithaca. New York State, pp. 231.

Foote RH. **1999b**. Development of reproductive biotechnologies in domestic animals from artificial insemination to cloning: a perspective. *Cloning*. **1**, 133-142.

Galli C, Duchi R, Crotti G and Lazarri G. **1998**. Embryo production by ovum pick up in water buffalo. *Theriogenology*. **49**, 400.

Gordon I. **1994**. Laboratory production of cattle embryos. CAB International Wallingford., UK, pp. 640.

Heape W. **1891**. Preliminary note on the transplantation and growth of mammalian ova within a uterine foster mother. *Proc. R. Soc. London*. **48**, 457-458.

Huang B, Cui K, Li T, Wang, X, Lu F, Liu Q, da Silva FM and Shi D. **2009**. Generation of buffalo (*Bubalus bubalis*) transgenic chimeric and nuclear transfer embryos using embryonic germ-like cells expressing enhanced green fluorescent protein. *Reprod. Domest. Anim.* (in press).

Huang B, Xie TS, Shi DS, LiT, Wang XL, MoY, Wang ZQ and Li MM. 2007. Isolation and characterization of EG-like cells from Chinese swamp buffalo (*Bubalus bubalis*). *Cell Biol. Int.* **31**, 1079–1088.

Hufana-Duran D, Pedro PB,Venturina HV, Hufana RD, Salazar AL, Duran PG and Cruz LC. **2004**. Post-warming hatching and birth of live calves following transfer of *in vitro*-derived vitrified water buffalo (*Bubalus bubalis*) embryos. *Theriogenology*. **61**, 1429–1439.

Ireland JJ and Roche JF. **1982**. Development of antral follicles in cattle after prostaglandin-induced luteolysis: changes in serum hormones, steroids in follicular fluid, and gonadotropin receptors. *Endocrinology*. **111**, 2077-2086.

Ireland JJ and Roche JF. **1983a**. Growth and differentiation of large antral follicles after spontaneous luteolysis in heifers: changes in concentration of hormones in follicular fluid and specific binding of gonadotropins to follicles. *J. Anim. Sci.* **57**, 157-67.

Ireland JJ and Roche JF. **1983b**. Development of non-ovulatory antral follicles in heifers: changes in steroids in follicular fluid and receptors for gonadotropins. *Endocrinology*. **112**, 150-156.

Ireland JJ and Roche JF. **1987**. Hypotheses regarding development of dominant follicles during a bovine oestrus cycle. Follicular growth and Ovulation rate in Farm Animals. J.F. Roche and D. O'Callaghan, eds. Martinus Nijhoff Publishers, The Netherlands, 1-18.

Jainudeen MR, Takahashi Y, Niyash M and Kanagawa H. **1993**. In vitro maturation and production of swamp buffao (*Bubalus bubalis*) oocytes. *Anim. Reprod. Sci.* **31**, 205-212.

Kamonpatana M. **1990**. Multiovulation in buffaloes leading to ET: the world's prospect and achievement at year 1990. *Buffalo J.* **6**, 1-10.

Kasiraj R, Misra AK, Rao MM, Jaiswal RS and Rangareddi NS. **1993**. Successful culmination of pregnancy and live birth following transfer of frozen-thawed buffalo embryos. *Theriogenology.* **39**, 1187–1192.

Kitiyanant Y, Saikhun J, Chaisalee B, White KL and Pavasuthipaisit K. **2001**. Somatic cell cloning in buffalo (*Bubalus bubalis*): effects of interspecies cytoplasmic recipients and activation procedures. *Cloning Stem Cells.* **3**, 97–104.

Laowtammathron C, Lorthongpanich C, Ketudat-Cairns M, Hochi S and Parnpai R. **2005.** Factors affecting cryosurvival of nuclear transferred bovine and swamp buffalo blastocysts: effects of hatching stage, linoleic acid–albumin in IVC medium and ficoll supplementation to vitrification solution. *Theriogenology.* **64**, 1185–1196.

Liang XW, Lu YQ, Chen MT, Zhang XF, Lu SS, Zhang M, Pang CY, Huang FX and Lu KH. **2008**. *In vitro* embryo production in buffalo (*Bubalus bubalis*) using sexed sperm and oocytes from ovum pick up. *Theriogenology.* **69**, 822–826.

Lu G and Hsu TT. **1991**. In vitro fertilization of oocytes in Taiwan water buffalo. In: *Proc. of 3rd World Buffalo Congress (Verna)*, **Vol. 3**, pp 604-609.

Lu YQ, Liang XW, Zhang M, Wang WL, Kitiyanan, Y, Lu SS, Meng B and Lu KH. **2007**. Birth of twins after *in vitro* fertilization with flow-cytometric sorted buffalo (*Bubalus bubalis*) sperm. *Anim. Reprod. Sci.* **100**, 192–196.

Madan ML, Singla SK, Jailkhani S and Ambrose JD. **1991**. *In vitro* fertilization in buffalo and birth of first ever IVF buffalo calf. **In**: *Proc. 3rd World Buffalo Congress, Varna.* Bulgaria, pp. 11–17.

Martin GR. **1981**. Isolation of pluripotent cell line from early mouse embryos cultured in medium conditioned by teratocarcinoma stem cells. *Proc. Natl. Acad. Sci. USA.* **78**, 7634-7638.

Neglia G, Gasparrini B, Caracciolo di Brienza, V, Di Palo R, Campanile G and Zicarelli L. **2003**. First pregnancies established from vitrified blastocysts entirely produced *in vitro* in Mediterranean Italian buffalo cows (*Bubalis bubalis*). *Theriogenology.* **59**, 374. [Abstract]

Nicholas FW. **1979**. The genetic implications of multiple ovulation and embryo transfer in small dairy herds. **In**: *Proc. 30th Annu. Meet. European Assoc. Anim. Prod. (Harrogate),* CG1.11.y

Parnpai R, Tasrpoo K and Kamonpatana M. **1999**. Development of cloned swamp buffalo embryos derived from fetal fibroblast: comparison of *in vitro* cultured with or without buffalo and cattle oviductal epithelial cells. *Buffalo J.* **15,** 371–384.

Pavasuthipaisit K, Kitiyananat Y, Thonabulsombat C, Tocharus C, Sriurairratna S and White KL. **1992**. In vitro maturation and fertilization of swamp buffalo oocytes and their subsequent development. *Theriogeneology*. **38,** 545-555.

Pierson RA and Ginther OJ. **1984**. Ultrasonography of the bovine ovary. *Theriogenology,* **21**, 495-504.

Presicce GA. **2007**. Reproduction in water buffalo. *Reprod. Domest. Anim.* **42** (Suppl. 2), 24–32.

Presicce GA, Rath D, Klinc P, Senatore EM and Pascale M. **2005b**. Buffalo calves born following AI with sexed semen. *Reprod. Domest. Anim.* **40,** 349.

Presicce GA, Verberckmoes S, Senatore EM and Rath D. **2005a**. First established pregnancies in Mediterranean Italian buffaloes (*Bubalus bubalis*) following deposition of sexed spermatozoa near the utero tubal junction. *Reprod. Domest. Anim.* **40,** 73–75.

Rajakoski, E. **1960**. The ovarian follicular system in sexually mature heifers with special reference to seasonal, cyclical and left-right variations. *Acta Endocrinol.* **34,** 7-68.

Rao KB, Pawshe CH and Totey SM. **1993**. Sex determination of *in vitro* developed buffalo (*Bubalus bubalis*) embryos by DNA amplification. *Mol. Reprod. Dev.* **36**, 291–296.

Shah RA, George A, Singh MK, Kumar D, Chauhan MS, Manik RS, Palta P and Singla SK. **2008**. Hand-made cloned buffalo (*Bubalus bubalis*) embryos: comparison of different media and culture systems. *Cloning Stem Cells.* **10,** 435–442.

Shah RA, George A, Singh MK, Kumar D, Anand T, Chauhan MS, Manik RS, Palta P and Singla SK. **2009**. Pregnancies established from handmade cloned blastocysts reconstructed using skin fibroblasts in buffalo (*Bubalus bubalis*). *Theriogenology*. **71**, 1215-1219.

Sharifuddin W and Jainudeen MR. 1988. Embryo collection in swamp buffalo (*Bubalus bubalis*). In: *Proc. 11th International Congress Animal Reproduction and AI (Dublin),* **vol. 4,** pp. 543

Sharma A, Datt M. Sharma A and Balakrishnan CR. **1994**. In vitro maturation, fertilization and cytogenetic evaluation of buffalo embryos- a review. *Indian J. Dairy Sci.* **47**, 714-722.

Shi D, Lu F, Wei Y, Cui K, Yang S, Wei J and Liu Q. **2007**. Buffalos (*Bubalus bubalis*) cloned by nuclear transfer of somatic cells. *Biol. Reprod.* **77**, 285-291.

Singh AK. **2006**. Studies on handmade cloning in buffaloes. M. V. Sc. Thesis submitted to Dairy Science College, National dairy Research Institute (Deemed University), Karnal. India.

Singh AK, Singh B, Singla SK. **2008**. Studies on handmade somatic cells cloning in buffalo (*Bubalus bubalis*). **In**: *International conference on emerging infectious diseases of animals and biotechnological applications*, held at Tamilnadu University of Veterinary and Animal Sciences, Chennai, India.

Singh B, Chauhan MS, Singla SK, Gautam SK, Verma V, Manik RS, Singh AK, Sodi M and Mukesh M. **2009**. Reproductive biotechniques in buffaloes (*Bubalus bubalis*): status, prospects and challenges. Review. *Reprod. Fertil. Dev.* **21**, 499-510.

Singla SK and Madan ML. **1990**. Response of superovulation in buffaloes (*Bubalus bubalis*) with Super-OV and FSH-P. *Theriogenology*. **33**, 327.

Singla SK, Manik RS and Madan ML. **1997**. Micromanipulation and cloning studies on buffalo oocytes and embryos using nucleus transfer. *Indian J. Exp. Biol.* **35**, 1273–1283.

Sritanaudomchai H, Pavasuthipaisit K, Kitiyanant Y, Kupradinun P, Mitalipov S and Kusamran T. **2007**. Characterization and multilineage differentiation of embryonic stem cells derived from a buffalo parthenogenetic embryo. *Mol. Reprod. Dev.* **74**, 1295–1302.

Suzuki T, Singla SK, Jailkhani S and Madan ML. **1992**. Cleavage capability of water buffalo follicular oocytes and their ability to cleave in vitro. *Theriogenology*. **38**, 1187-1194.

Techakumphu M, Lohachit C, Chantaraprateep P, Prateet P and Kobayashi G. **1989**. Preliminary report on cryopreservation of Thai swamp buffalo embryos: manual and automatic methods. *Buffalo Bull.* **8**, 29–36.

Totey SM, Taneja M, Pawshe CH, Singh GP and Talwar GP. **1991**. In vitro maturation, fertilization and development of buffalo oocytes. **In**: *Proc. 3rd World Buffalo Congress (Verna)*, **Vol. 3**. pp. 610-617.

Totey SM, Pawshe CH and Singh GP. **1993a**. In vitro maturation and fertilization of buffalo oocytes (*Bubalus bubalis*): effects of media, hormones and sera. *Theriogenology*. **39**, 1153-1171.

Totey SM, Pawshe CH, Singh GP. **1993b**. Effect of bull and heparin and sperm concentration on in vitro fertilization of buffalo (*Bubalus bubalis*) oocytes matured in vitro. *Theriogeneology*. **39**, 887-898.

Verma V, Gautam SK, Singh B, Manik RS, Palta P, Singla SK, Goswami SL and Chauhan MS. **2007**. Isolation and characterization of embryonic stem cell-like cells from *in vitro*-produced buffalo (*Bubalus bubalis*) embryos. *Mol. Reprod. Dev.* **74**, 520–529.

Verma V, Gautam SK, Palta P, Manik RS, Singla SK and Chauhan MS. **2008**. Development of a pronuclear DNA microinjection technique for production of green fluorescent protein-expressing bubaline (*Bubalus bubalis*) embryos. *Theriogenology*. **69**, 655–665.

Wadhwa N, Kunj N, Tiwari S, Saraiya M and Majumdar SS. **2009**. Optimization of embryo culture conditions for increasing efficiency of cloning in buffalo (*Bubalus bubalis*) and generation of transgenic embryos via cloning. *Cloning Stem Cells*. **11**, 387-395.

Waewick BL and Berry RO. **1949**. Intergeneric and intraspecific embryo transfers in sheep and goat. J. Hereditary. 40, 297-303.

Whittingham DG. **1971**. Survival of mouse embryos after freezing and thawing. *Nature*. **233**, 125-126.

Willet EL, Black WG, Casida LE, Stone WH and Buckner PJ. **1951**. Successful transplantation of a fertilized bovine ovum. *Science*. **113**, 247.

Wilmut I, Rowson LE. **1973**. Experiments on the low temperature preservation of cow embryos. *Vet. Rec.* **92**, 686-690.

Wilmut I, Schnieke AE, McWhir J, Kind AJ and Campbell KH. **1997**. Viable offspring derived from fetal and adult mammalian cells. *Nature*. **385**, 810- 813.

—□□□—

3

Semen Technology in Assisted Reproduction

"Semen production and trade is a worldwide industry in livestock production sector. In commercial livestock breeding, artificial insemination has a major role in terms of health and economic production. The most important challenge is, therefore, the availability of high quality semen from a progeny tested bull"

Introduction

Sperm is essential for contributing at least three components: (i) the paternal haploid genome, (ii) induction of molecular signals to initiate activation of the oocyte, and (iii) the centriole, which directs microtubule assembly leading to the formation of the mitotic spindles during the initial zygote development (Barroso et al., 2009). Past research has viewed the spermatozoon as a simple carrier that transfers paternal DNA to the egg and playing some role in embryogenesis. It is now well established that there is extensive cross-talk between the fertilizing-sperm and the egg, leading to activation of the egg on one hand and sperm-head decondensation on the other. This is orderly followed by female and male pronuclear formation, syngamy, and the first cleavage divisions (Barroso et al., 2009). Together these molecular events and processes are integral components of sexual reproduction and fertility in mammalian species.

Like other livestock species, the male infertility in buffalo bull is also influenced by genetic and environmental factors. The sexual maturity of male buffaloes is attained at about 2-3 years of age depending on type of breed and management practices. Sharma (2003) have reported that age of the bull has significant effect on desirable semen characteristics like semen volume, sperm concentration and sperm motility. To an extent winter months are more favorable for attaining early sexual maturity in buffaloes.

Similarly, successful embryo development depends much on genetic and epigenetic contributions from both the male and female. This presents opportunities for identifying biological markers for spermatogenic damage, as well as protective measures (Chenoweth, 2007).

Spermatogenesis in buffalo bulls

During the process of spermiogenesis in mammals, a significant reorganization of the chromatin structure occurs involving the sequential substitution of nuclear histones with protamines. Spermatogenesis in buffalo-bulls is an orderly and well defined process, occurring in the testes, where stem cell spermatogonia at the base of seminiferous tubules divide mitotically to maintain their own numbers and to cyclically produce the spermatocytes. The spermatocytes undergo meiosis to produce haploid spermatids, which differentiate into spermatozoa that are released into the testicular lumen. The three major divisions of spermatogenesis, which are characterized by the development of spermatogonia (spermatocytogenesis), spermatocytes (meiosis) and spermatids (spermiogenesis), each occupy one-third of the total duration of sperm production (Gordon, 2003). Cellular elements of the seminiferous tubules, namely, spermatogonia, primary spermatocytes, secondary spermatocytes, spermatids etc. are recognized on the basis of their shape, size, chromatin pattern and nucleolus topography. Spermatogenesis in buffalo involves division and transformation of three morphologically different types of spermatogonia i.e. type-A, intermediate and type-B from spermatogonial stem cells. This is unlike rat and mice where more than five different types of spermatogonia are involved in the spermatocytogenesis. Last spermatogonial division gives rise to primary spermatocytes which subsequently undergo two distinct meiotic divisions and produce spermatids.

In buffalo-bulls, the spermatogenesis is divided into eight distinct stages. These eight stages can be clearly identified on the basis of morphology and location of different cell types. In stage-1, the spermatids lying towards the lumen become rounded in shape. Then become oval in stage-2, and completely elongated and thread like in stage-3. First and second meiotic divisions take place in stages-4 and stage-5, respectively. Stage-6 and stage-7 of the seminiferous epithelial cycle are identifiable from the centripetal movement of the elongated spermatids. In stage-8 the fully formed spermatozoa tend to line the lumen of seminiferous tubules.

Various factors which influence the sperm production efficiency include the number of spermatids taking part in the process of spermiogenesis, diameter of the seminiferous tubules, length of the seminiferous tubules, weight of testicular tissue parenchyma and storage capacity of epididymis, particularly the cauda epididymis.

Assessing bull-fertility

It is important to select sires with a high fertility and breeding soundness which refers to a bull's ability to get the females pregnant. A standard breeding soundness evaluation identifies bulls with substantial fertility, but does not consistently identify sub-fertile bulls. Although the fertility is a genetically- and environmentally-controlled trait, the overall reproductive efficiency to a large extent can be controlled through managerial practices like feeding, exercising and general care of the bulls. Whereas the poor semen-quality is an obvious indicator of sub-fertility, good semen may also not always guarantee acceptable fertility (Den Daas, 1992). This is because the spermatozoa of different bulls may vary in their ability to fertilize oocytes. For this reason, considerable effort is being invested in identifying the markers for functional capacity of the sperm and thereby accurate prediction of fertility of the sire.

In the past, selection of a bull was often made on phenotypic characteristics. Even today, despite vast information available on semen analysis for the farm animals, no single test is found suitable to meet the needs or criteria involved in AI or fertility-merit of a bull. Traditionally, the bull semen-quality is judged on semen wave motion and progressive sperm motility and an objective evaluation based on sperm concentration and morphological abnormalities (Gordon, 2003). Moreover, evaluation of male-fertility by mating or AI is expensive and time-consuming, and only allows a limited number of males to be tested. Therefore, development of an *in vitro* assay to determine bull-fertility would result in considerable cost savings compared to expensive field-based fertility testing trials. In this respect, interest has been focused on test methods related to IVF techniques. Assays based on the ability of spermatozoa to bind to homologous or heterologous zona pellucida (ZP) have been developed as a diagnostic test in several animal species (Ivanova and Mollova, 1993; Liu and Baker, 2000), and formation of pronuclei after IVF (Songsasen and Apimeteetumrong, 2002).

During fertilization in mammals, sperm must bind to and penetrate the ZP to reach, bind to and fuse with the egg plasma membrane. Considering this as a criterion, a recent study was aimed to develop a system of *in vitro* assays based on ZP-binding and IVF for predicting male fertility in buffalo bulls, and it was shown that the application of IVF technology has potential in fertility selection of buffalo bulls, and therefore, provides a basis to predict and maximize the IVF performance of bulls (Yadav *et al.*, 2008; Abdel Dayem *et al.*, 2009). In the study, no. of sperm attached per ova was taken as criterion for assessing the fertility of buffalo bulls; higher attachment leading to higher fertilization rate.

Biochemical attributes of buffalo semen

Basic information on buffalo semen constituents is essential for the development of semen extenders for dilution and preservation of sperm. The spermatozoa of the buffalo bull are morphologically different from those of the cattle bull which are shorter and narrower. The average measurement is as follows: head length is 8 microns and breadth is 5 microns mid piece length is 15.5 microns and tail length 55 microns. The normal color of the semen is opaque, milky white like that of a cattle bull. The average semen volume for Indian buffalo is 2-3 cc per ejaculation while in Murrah buffalo bulls it is 3-4 cc per ejaculation.

The plasma membrane of buffalo spermatozoa contains higher level of polyunsaturated fatty acid (Jain and Anand, 1976). Besides cholesterol, seminal plasma also contains triacylglycerols which are metabolized to meet the sperm energy requirements. The concentration of ascorbic acid in the buffalo semen has been reported to be low and its supplementation during sperm cryopreservation is reported to be useful (Sidhu and Guraya,1985). Cragle and Salisbury (1959) showed that potassium levels in extending medium, comparable to those found in the seminal plasma, had an inhibitory effect on oxygen-consumption, fructose utilization and lactic acid accumulation.

Semen processing: Storage, thawing and handling

Several criteria and procedural steps are involved in determining the efficient preservation and utilization of the semen. Choice of diluter or extender is important. Semen extenders are mainly based either on milk or on egg yolk. Extenders with a composition similar to the original recipe formulated by Kenney *et al.* (1975) are most popular and used worldwide with suitable modifications. Recently, numerous defined media have been developed to minimize or exclude non-desirable components and their detrimental effects on spermatozoa. These extenders are inexpensive, easy to prepare, can be stored in frozen form and result in acceptable fertility rates. Egg yolk (EY), the important ingredient of semen extenders protects cell membranes against cold shock, and it prevents or restores the loss of phospholipids from the sperm membrane. Important implications of AI, its limitations and properties of semen extendes have been ennumerated in **Table 1-2** and **Table-3** summarizes the important attributes of semen extenders.

As noted earlier the season appears to affect bull fertility significantly. A poor cryosurvival of semen with a higher rate (up to 50%) of backward-moving spermatozoa was observed during April to July which resulted in higher rejection of buffalo semen by the farmers or buffalo breeders. A study on effect of osmolarity and stage of glycerolization of semen extender with reference to their effects on buffalo sperm have revealed that time of addition of glycerol had critical impact on sperm-motility. Further, addition of glycerol at initial stages of semen dilution (room temp.) instead of at 5°C could overcome other problem of backward motility of spermatozoa during cryopreservation observed in summer season. It has been observed that glycerolization at initial stage (room temp.) of dilution could overcome the problem of impaired motility in spermatozoa subjected to cryopreservation (Singh P, CIRB Hisar, personal communication).

Buffalo bull semen preservation

Important function of a preserving medium is to increase the total volume of usable semen and to maintain the fertility of spermatozoa. The conventional extender like egg yolk citrate (EYC) used for bull semen preservation was found to be partially effective for buffalo bull semen. The use of some other extenders like egg yolk glucose sodium bicarbonate (EYG-SB), skim milk egg yolk (SMY) and coconut milk extender (CME) have resulted in some success for shorter-term preservation but were not suitable for deep freezing of semen. The use of buffers like tris (trihydroxymethyl-amino methane) citrate for cryopreservation of buffalo bull semen has given encouraging results (Sengupta and Chaube, 1972). Ganguli *et al.* (1973) reported successful preservation of buffalo semen in citric acid whey (CAW) extender and obtained 50% conception rate in 297 inseminations with the semen stored at 5-7°C for 72 h. Fertility trials with refrigerated buffalo semen diluted in tris buffer based extender and citrate glycine extender and did not reveal any significant difference in conception rate by using two extenders, although the conception rate was slightly higher (65.0%) in citrate glycine than in tris buffer based medium (59.7%) (Chaube and Sengupta, 1974). Tris citric acid buffer based extender has been used for deep-freezing as well as liquid preservation of buffalo semen in most of the semen laboratories in India and abroad.

Advances in male reproduction technology

Sperm sexing

Ever since the invention of flow-cytometry (Kamentsky and Melamed, 1967) and utilization of DNA as a marker for the selection of X- and Y-bearing sperm (Moruzzi, 1979), the applications of male germplasm has gained new directions for production of desired offspring. At the moment flow-cytometric-sorting system has improved greatly and successful cryopreservation of sex-sorted bull sperm has promoted the commercialization of sexed-semen globally. Additionally, the sperm-sexing has raised the possibility of improving livestock production through selective production of female or male calves. Initially a sperm-sexing system could only sort about 400,000 sperm per hour. Thus, it would take 25 h to sort one insemination dose of 10×10^6 sperm of each sex, but has now improved to meet the commercial demands. Sperm sexing is still in initial phases of experimentation in buffaloes. This aspect has also been discussed in the chapter on sex-preselection.

Microencapsulation of sperm

A variety of methods are known by which biological material can be encapsulated in the form of microcapsules. In the processes the material to be encapsulated is generally dispersed in a solvent containing a wall forming material. At a single stage of the process, solvent is removed from the microcapsules and thereafter the microcapsule product is obtained. Microencapsulation of sperm has been in use in bovines for about two decades. Nebel *et al.* (1993), had originally reported a technique for microencapsulation of bovine spermatozoa with minimal spermatozoa-injury and thus of potential use in AI. The polymers comprised of poly-l-lysine, polyvinylamine and protamine sulfate were proven to be suitable for encapsulation membranes (Nebel *et al.*, 1993). Uterine retention of inseminates was favored by capsules having a 'sticky' membrane. Using these procedures, preliminary homospermic fertility studies have indicated that sperm encapsulated with poly-l-lysine or protamine sulfate may achieve normal fertility (Nebel *et al.*, 1993).

A technology that permits the insemination event to be dissociated from the ovulatory event has many potential applications. For example, in animals under free-range conditions, a single mustering would be immensely beneficial and would provide the genetic enhancement of the livestock. In countries with limited technological capabilities, an insemination team equipped with frozen-encapsulated spermatozoa from a technology centre might be able to reach personally the remote areas and inseminate herd thus, providing the economic utilization of technology to a wider market. In many non-domestic species, oestrus is not well defined. To be able to insert a micocapsule independently of the ovulatory event would offer advantages in increasing the pregnancies in such species. Microencapsulation procedures are compatible with sperm viability and can be adapted to an acceptable extender system used in AI. However, there is much yet to be investigated about the larger scale use of this technology.

Spermatozoa from stem cells

Production of embryonic stem (ES) cells in important livestock species and generation of functional germ cells from these ES cells in sufficient number will have important consequences for infertility treatment, culture of germ cell precursors, breeding of livestock species and conservation of endangered species. This is a newly emerging field of assisted reproduction, and has been discussed in chapter on stem cells technology.

Emerging trends in semen technology

Assessing semen quality is conventionally based on macroscopic (volume, color, odor, and consistency) and microscopic (motility, morphologic properties, and concentration) analyses (Beer-Ljubic *et al.*, 2009). However, studies have shown such assessment of criteria to be inadequate, as semen declared as 'high-quality' by conventional analysis may have reduced fertilizing potential. Therefore, determination of additional biochemical parameters has been proposed (Meseguer *et al.*, 2004; Argov *et al.*, 2007; Brinsko *et al.*, 2007; Turba *et al.*, 2007). This is because sperm structure and fertilizing potential depend on lipid composition in sperm cell membrane and seminal plasma. Ovum fertilization is preceded by the process of capacitation that is accompanied by modifications in the sperm membrane lipid composition. These changes imply the significance of membrane lipids in the sperm membrane and binding to protein in seminal plasma and female reproductive organs during *in vivo* fertilization (Therien *et al.*, 1998; Witte and Schafer-Somi, 2007).

Opportunities and challenges

In comparison to cattle, the buffalo presents no specific problems for the application of the technique of AI. It is generally believed that the keeping quality of the buffalo semen is lower than that of the bull semen and the research on the keeping quality of buffalo semen, its storage and techniques of AI is in progress in India.

Looking at the rapid development in the recent past, production of sperm cells and oocytes from alternate sources holds vast scope for future applications in human beings and livestock species. Spermatogenesis and oogenesis in culture should contribute to various areas, including advance studies on fertility treatment and germ and somatic cell interaction. This shall increase our understanding regarding genetic control of gamete formation. Future perspectives of germ cells production from alternate route can be utilized in endangered animals, besides vast scope in cell therapy and developmental biology studies. Transgenic germ cells with desired traits hold the key for harvesting the maximum potential from this promising technology.

Conclusion

Sperm has a very dynamic and critical participation in normal embryogenesis that clearly extends beyond the fertilization process. Clinical evidences clearly explain that defective spermatozoa fertilizing the oocyte may cause arrest of development at multiple levels during embryo preimplantation development. Additional data suggest that sublethal effects can be "carried over" after implantation, resulting in untoward embryonic/ fetal defects. Demand and effectiveness of buffalo semen can be enhanced by optimizing the protocols for processing, storage and transport of semen, the use of proven bulls and the improvement in AI technology and policies.

Table 1. Applications of artificial insemination in livestock production and reproduction management

1. Increasing the usefulness of superior sire to an extra ordinary degree. Semen can be collected from high value aged bulls and processed for cryopreservation or insemination.
2. Extending the services of superior sires. By natural services, a bull can be bred to 50 to 60 females whereas by AI the semen from a single ejaculate can be used to inseminate hundreds of females

3. Avoiding the necessity to maintain a herd sire and thus avoiding the bottlenecks accompanied with the management (searching and purchasing) of new breeding bull
4. AI helps in regulating the breeding program and reducing the interval between successive calving without unnecessarily prolonging the dry period
5. Exploiting the cross breeding strategies for hybrid vigor by quickly transporting the semen globally
6. Minimizing the intensity of spread of genital diseases, increasing the rate of conception by conducting AI under complete sanitary conditions by the specially trained persons.

Table 2. Limitations of artificial insemination technique

1. Requirement of a technically skilled person, and sound knowledge of anatomy of female reproductive tract
2. AI requires expensive consumables and equipments, insanitary conditions and skilled person to use the instruments
3. Requirements of more than one insemination compared to natural services
4. Improper cleaning of the instruments may lead to spread of microbial infections and reduced fertility.
5. AI needs a scientifically validated selection of the sire
6. Preservation and transportation of semen is costly and difficult under severe climatic conditions like those prevailing in most parts of developing countries.

Table 3. Properties of semen diluters

1. To be isotonic with semen
2. To provide buffering capacity during the sperm handling and storage
3. Protecting the sperm from cold shock during cooling from body temperature to low- and ultralow-temperature
4. Resistance to microbial growth
5. Preservation of sperm life and minimum drop in fertility

References

Abdel Dayem AMH, Mahmoud KM, Nawito MF, Ayoub MM and Scholkamy TH. **2009**. Fertility evaluation in Egyptian buffalo bulls using zona pellucida binding and in vitro fertilization assays. *Livestock Sci.* **122**, 193-198.

Argov N, Sklan D, Zeron Y and Roth Z. **2007**. Association between seasonal changes in fatty-acid composition, expression of VLDL receptor and bovine sperm quality. *Theriogenology*. **67**, 878–885.

Barroso G Valdespin C, Vega E, Kershenovich R, Avila R, Avendano C and Oehninger S. **2009**. Developmental sperm contributions: fertilization and beyond. *Fertil. Steril*. **92**, 835-847.

Beer-Ljubic, B, Aladrovic, J. Marenjak TS, Laskaj R, Majic-Balic, I and Milinkovic-Tur S. **2009**. Cholesterol concentration in seminal plasma as a predictive tool for quality semen evaluation. *Theriogenology*. **72**, 1132-1140.

Brinsko SP, Love CC, Bauer JE, Macpherson ML and Varner DD. **2007**. Cholesterol-to-phospholipid ratio in whole sperm and seminal plasma from fertile stallions and stallions with unexplained subfertility. *Anim Reprod Sci*. **99**, 65–71.

Chaube, LK and Sengupta, BP. **1974**. A note on the fertility of buffalo semen in Tris diluent. *Indian Vet. J*. **51**, 388.

Chenoweth PJ. **2007**. Influence of the male on embryo quality. *Theriogenology*. **68**, 308-315.

Cragle RG and Salisbury GW. **1959**. Factors influencing metabolic activity of bull spermatozoa. IV pH, osmotic pressure, and the cations, sodium, potassium and calcium. *J. Dairy Sci*. **42**, 1304.

Den Daas N. **1992**. Laboratory assessment of semen characteristics. *Anim. Reprod. Sci*. **28**, 87–94.

Ganguli, NC Bhosrekar, M and Stephan Jose. **1973**. Milk Whey as a diluent for buffalo semen. *J. Reprod. Fert*. **35**, 355.

Gordon I. **2003**. Laboratory Production of Cattle Embryos. Gordon, I (Ed.), 2nd Edition. CABI Publishing, CAB International Wallingford, Oxon OX10 8DE, UK.

Ivanova M. and Mollova M. **1993**. Zona-penetration in vitro test for evaluating boar sperm fertility. *Theriogenology*. **40**, 397–410.

Jain YC and Anand SR. **1976**. The lipids of buffalo spermatozoa and seminal plasma. *J. Reprod. Fertil*. **47**, 255.

Kenney RM, Bergman RV, Cooper WL and Morse GW. **1975**. Minimal contamination techniques for breeding mares: techniques and preliminary findings. **In**: *Proc. 21st Annu. Conv. Am. Assoc, Equine Pract*, p. 327.

Kamentsky LA and Melamed MR. 1967. Spectrophotometric cell sorter. *Science*, **156**, 1364–1365.

Liu DY and Baker HWG. **2000**. Defective sperm-zona pellucida interaction: a major cause of failure of fertilization in clinical in-vitro fertilization. *Hum. Reprod.* **15**, 702–708.

Meseguer M, Garrido N, Martinez-Conejero JA, Simon C, Pellicer A and Remohi J. **2004**. Relationship between standard semen parameters, calcium, cholesterol content and mitochondrial activity in ejaculated spermatozoa from fertile and infertile males. *J. Assist. Reprod. Genet.* **21**, 445–451.

Moruzzi JF. **1979**. Selecting a mammalian species for the separation of X- and Y-chromosome-bearing spermatozoa. *J. Reprod. Fertil.* **57**, 319-323.

Nebel RL, Vishwanath R, McMillan WH and Saacke RG. **1993**. Microencapsulation of bovine spermatozoa for use in artificial insemination: a review. *Reprod. Fertil. Dev.* **5**, 701-712.

Sharma AK. **2003**. Spermatogenesis in buffaloes and its practical implications. **In**: *Summer School on Artificial Breeding and Reproduction Management in Buffaloes,* organized by CIRB Hisar, June 10-30, 2003.

Sidhu KS and Guraya SS. **1978**. Effect of cold shock on enzyme release in buffalo spermatozoa. *Ann. Biol. Anim. Bioch. Biophys.* **18**, 283.

Sengupta BP and Chaube LK. **1972**. Preservation of buffalo semen in tris buffer-based diluents. *Indian J. Anim. Sci.* **42**, 165.

Songsasen N. and Apimeteetumrong M. **2002**. Effects of β-mercaptoethanol on formation of pronuclei and developmental competence of swamp buffalo oocytes. *Anim. Reprod. Sci.* **71**, 193–202.

Therien I, Moreau R and Manjunath P. **1998**. Major proteins of bovine seminal plasma and high-density lipoprotein induce cholesterol efflux from epididymal sperm. *Biol. Reprod.* **59**, 768–776.

Turba ME, Fantinati P, Bernardini C, Gentilini F, Bacci ML and Forni M. **2007**. Relationships between innovative and traditional parameters to investigate semen quality in pigs. *Anim. Reprod. Sci.* **99**, 72–81.

Witte TS and Schafer-Somi S. **2007**. Involvement of cholesterol, calcium and progesterone in the induction of capacitation and acrosome reaction of mammalian spermatozoa. *Anim. Reprod. Sci.***102**, 181–193.

Yadav PS, Singh I. Sethi RK. **2008.** *In vitro* cleavage rates of homologous oocytes for fertility evaluation of buffalo bulls. *Indian Vet. J.* **7**, 725-727.

—□□□—

4

Embryo Production and Transfer

"In a little over a century, embryo biotechnology has changed beyond all recognition and speculations and receiving a lot of publicity in the media and scientific community. Apart from conventional dissemination of elite genotypes, the embryo biotechnology is immensely useful in the production of embryonic stem cells and transgenic animals"

Introduction

Reproduction is the backbone of a commercial livestock production system. Embryo production and transfer technology in scientifically managed herds offers the potential for greatly extending the range of genetic material that may be conserved for future applications **(Table 1).** The technology is focused on augmenting *in vitro* production of embryos, micromanipulation of the embryos (splitting, nuclear transfer, embryonic stem cell derivation etc.) and transfer of high quality embryos **(Fig. 1)** to establish high yielding herds. The technology offers enormous opportunities for the livestock researchers and breeders to optimize genetic progression and increase the number of offspring born from different bloodlines within a timeframe. The embryo transfer (ET) offers the veterinarians and livestock breeders multiple opportunities to extend livestock health planning from the core ET group to rest of the herd, therefore, demonstrating that they can play a valuable role as part of the herd management team (Smith, 2009).

In a natural course, a single female will produce a single progeny in a year, but using superovulation and embryo transfer, it is possible to collect around 36 embryos from a single female in one year. Assuming an average success rate of 50% in the ET, an average of 18 progenies can be derived from one superior female in one year. Production of healthy, genetically superior and transferable quality embryo banks is, thus, the need of the hour. This will ensure birth of healthy and high-producing progeny. Besides its use in livestock production, the embryo production technology is an integral component of studying the reproduction biology in mammalian species, as evidenced by the recent exponential increase in the literature on the subject.

Principles of the technology

The commercial embryo production involves superovulation of the selected females, insemination or naturally serving the female to fertilize the multiple ova shed or *in vitro* fertilization (IVF) of the oocytes recovered from the live or slaughtered animals and matured *in vitro*. The technology once relied much on superovulation of the females and production of transferable quality multiple embryos. Embryo transfer can also be used for some specialized purposes such as rapid expansion of rare genetic stocks, reducing the cost of international transport by shipping embryos rather than live animals, the rapid replacement of existing genotypes by using ET rather than grading up through continuous crossing, and the possibility of increasing the twinning rate.

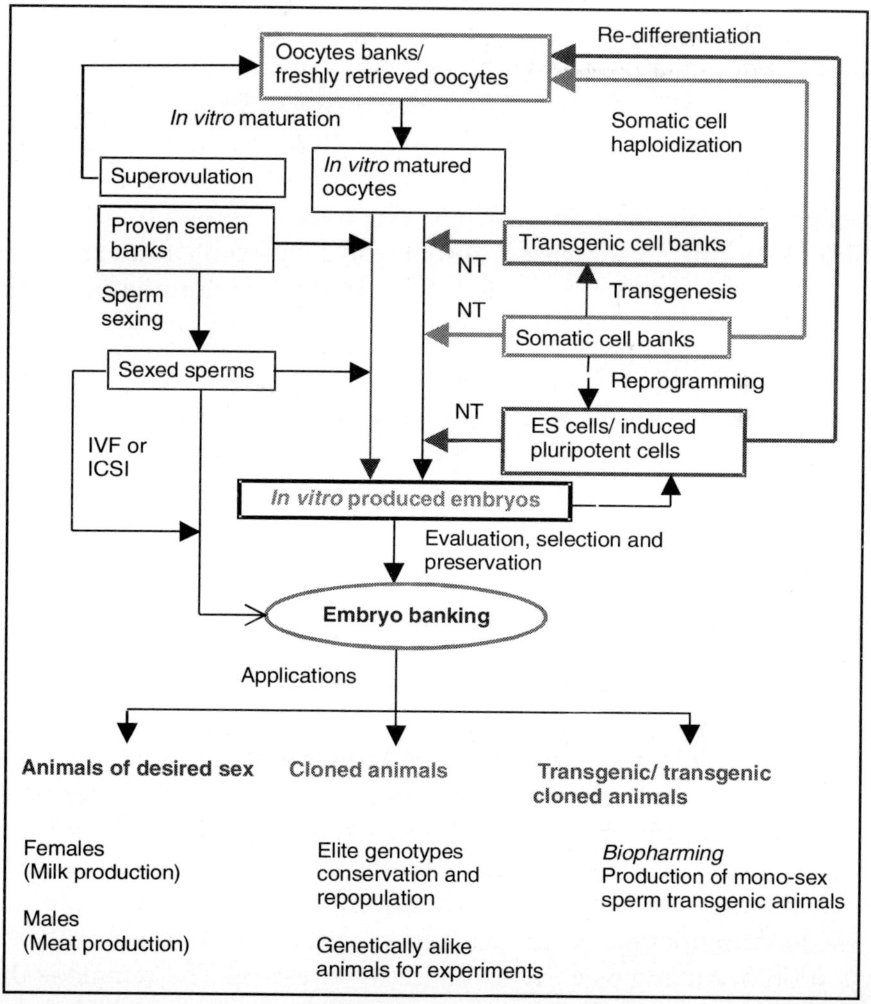

Fig. 1. Assisted reproduction and genetic engineering tools in livestock production induced pluripotency, transgenesis, and *in vitro* derivation of gametes are still under investigation in buffaloes.

Superovulation

Although ET has been in use in research since 1930s, originally with laboratory animal species such as rabbits or the mouse, it was in cattle that the technique was for the first initiated in livestock farms. The technique was initially used to facilitate the introduction and dissemination of beef cattle breeds. The emphasis then changed from beef cattle to dairy cattle and the uptake of the technology was largely in the hands of a generation of producers who had made good use of AI in dairy breeding programs (Gordon, 2003).

The multiple ovulation and embryo transfer (MOET), in combination with AI was initiated as a new breeding strategy in cattle production (Nicholas, 1979) has now undergone a rapid metamorphosis for applications in other livestock species. The term "superovulation" refers to induced stimulation of ovary through exogenous administration of drugs like, follicle stimulating hormone (FSH) to recruit an increased number of developing follicles on the ovary by overcoming the endocrine suppression of dominating follicle and continuing the follicular development to ovulation with concomitant maturation of the developing oocytes.

In principle, MOET type nucleus breeding schemes should have significant advantages in developing countries where the absence of on-farm recording makes conventional progeny testing impossible. This is particularly true for milking buffalo populations which do not have the option (like in cattle) of cross-breeding with highly selected breeds. The National Dairy Development Board (NDDB) has been involved since 1986 in developing such an ET-based scheme for buffalo.

The superovulation protocols in buffaloes relied mostly on recipes that were originally applied for cattle (Singla and Madan, 1990). For this, a schedule comprising of twice-daily administration of the FSH over the four days commencing mid-cycle (days 9-14) is followed. Incorporated in this regime is a prostaglandin-F2α (PGF2α) injection which is often given alongside the fifth FSH injection, to return the donor animals to estrus (Smith, 2009). The onset of estrus and subsequent ovulation occurs usually after the PGF2α treatment, with ovulation occurring as early as the onset of estrus. The female is then inseminated for getting the oocytes fertilized. The regime of drug administration and insemination may vary depending on the genetic, nutritional or health status of the animal.

Embryo collection and evaluation

It is performed generally on day 7 of the estrus cycle (day 0 = day of estrus). The developing embryos migrate to the lumen from the oviduct on approximately day 4 of the estrus cycle. The embryos collected at this stage are more appropriately suitable for transfer as well as cryopreservation. In larger ungulates including cattle and buffalo the collection is non-surgical requiring catheterization and flushing of the uterine horns. A two-way catheterization is used in cattle and buffaloes. Composition of the flushing media also varies among the different laboratories, but Dulbecco's phosphate buffered saline (DPBS) fortified with bovine serum albumin (BSA) and antibiotics is generally used.

Once the uterine horns are flushed, the embryos are searched in the flushed medium. The contents are allowed to settle down and then searched under stereo zoom microscope. A magnification of X6 to X10 is generally suitable for searching the bubaline embryos. The embryos are then held in appropriate medium for grading. Different laboratories employ their own criteria for grading of the embryos. However, as a general consensus, the embryos with a compact and uniform cellular integrity, intact zona pellucida and homogeneous cellular granulation are the qualifying features for an embryo to be of high quality.

Embryo freezing or transfer

The option for freezing the embryo or transferring it depends on the availability of estrus-synchronized surrogate mother, or the objective of the experiment. The surplus embryos can be cryopreserved and used at later stages.

In vitro production of embryos

It is a multiple step process involving recovery of oocytes, maturation and fertilization *in vitro*, embryo culture, evaluation and transfer or cryopreservation **(Fig. 1)**.

Oocyte recovery

There are different ways to derive oocytes and the oocyte quality is highly variable in different methods. The choice of the method depends on availability of the donor, technical skills and facilities available to collect the oocytes. Collection of the oocytes is generally performed by the isolation of ovarian follicles **(Fig. 2)** dissection, slicing or aspiration or oviductal flushing. The recovery technique and the quality of the oocytes collected depend on whether collection is made from slaughtered animals or from live animal, or at repeated occasions if to be collected from live animals.

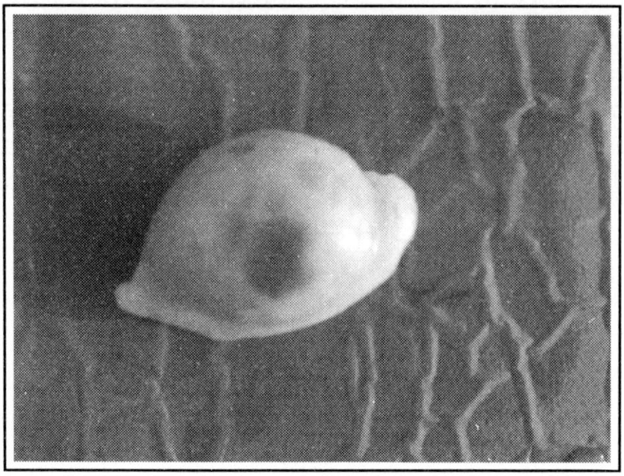

Fig. 2 Ovarian Follicle

Oocytes from live animals

In dealing with live animals, the oocytes are obtained from the surgically removed ovaries (ovariectomy) or by the laparoscopy-guided oocyte pick up (OPU). This technique offers several distinctive advantages **(Table-2).** Considerations are also for removing the ovaries prior to animal slaughter (Alberio *et al.*, 1998). Ovariectomy may enable getting oocytes in better hygienic conditions, and more reliable identification of the oocytes than those collected from the ovaries from abattoirs. However, the disadvantages include requirement of expensive equipments and technically skilled personnel. Also, a limited availability of donor animals and the poor-quality oocytes recovered from slaughtered animals discourage the larger scale applications of the technique.

Oocytes from abattoir-derived ovaries

Owing to the demand of the technical expertise, and cost involved in the recovery of ovaries or oocytes from live animals, the slaughter house derived ovaries **(Fig. 3)** are generally considered as cheaper and easily accessible sources of oocytes. It is, therefore, necessary to develop methods that facilitate the recovery of good quality oocytes per ovary. A review by Nagai (2001) addresses the criteria of selection of the ovaries likely to yield competent oocytes from animals at abattoirs.

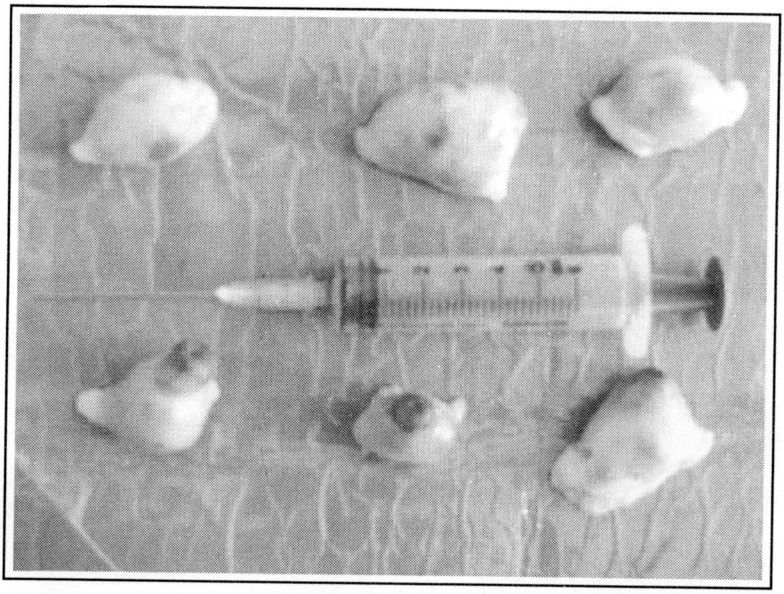

Fig. 3. Abattoir ovaries

The methods used for recovering the oocytes for embryo production may differ according to their source. The ovaries collected from the abattoir need a due care and attention during their transportation and handling. The buffalo ovaries are collected from the animals immediately after their slaughter, and washed with an isotonic sterile saline supplemented with a combination of antibiotics (400 IU mL^{-1} penicillin and 500 mg mL^{-1} streptomycin). A review by Sirard and Blondin (1996) has noted the evidence showing that developmental competence of bovine oocytes may be enhanced by exposure of the ovaries to warm incubation for a few hours before aspirating the oocytes. However, there is no defined temperature for maintenance of the animal ovaries prior to follicular aspiration. In buffaloes also, a temperature of 32°C-37°C may be a general recommendation, but it is practically difficult to maintain a uniform ambient temperature particularly when ovaries are to be transported across the longer road distances. The ovaries held at this temperature can yield good quality oocytes which can be used for *in vitro* production of embryos including nuclear transfer cloned embryos.

Aspiration of the oocytes

The techniques optimized in cattle can usually be attempted, albeit less successfully, in water buffaloes. The antral follicles are aspirated by using various devices such as syringe and needles, aspiration needle under vacuum pressure **(Fig. 4)**, and micropipette etc. One of the big problems in aspiration of oocytes from the buffalo ovaries is that some follicles remain embedded in the ovaries. Hence, the surface visible (3-8 mm diameter) as well as embedded oocytes have to be aspirated using hypodermic needle attached to high quality 10-20 ml disposable plastic syringes. In buffaloes, use of 18-19 gauge needles attached to 10-20 ml disposable syringes have been found suitable to aspirate a large number of ovaries in single batches. The oocytes are collected in aspiration medium comprising of Medium-199 fortified with BSA and antibiotics. The major advantage of follicle aspiration is in terms of speed of aspiration, which may be important in commercial embryo production system.

Fig. 4. Aspiration of oocytes

Alternatively, the oocytes can also be collected by dissection of ovaries or slicing the ovaries into smaller pieces in aspiration medium, removing the sliced tissue, and then searching the oocytes. Several studies have been carried out on efficiency and quality of oocytes recovered by dissection and slicing of bovine ovaries. The method offers the advantage of obtaining the oocytes from the embedded follicles. Reports from India (Jain *et al.*, 1995; Kumar *et al.*, 1997) showed that slicing was the most effective of the three recovery methods employed with the slaughtered buffalo ovaries. However, the technique is more useful in sheep and goats where size of the ovaries is smaller.

Aspiration of the oocytes, choice of aspiration medium, evaluation and selection of the oocytes for *in vitro* maturation (IVM) and fertilization (IVF) play critical role in further developmental stages of the embryos. Generally, the oocytes with a compact multilayered (>4-5 layers) cumulus cells complex (COC) with homogeneous ooplasm are considered to be A-grade oocytes, and yield good quality embryos. For evaluation of the oocytes, 200X magnification is enough.

Assessing oocyte quality

There are different systems used in the grading of buffalo oocytes. Most depend on visual assessment of the morphological features by the experienced personnel; hence can vary with the individual and the laboratory. The assessment criteria for buffaloes as suggested by Madan *et al.*, (1994) and Chauhan *et al.*, (1998), are primarily based on the presence of compact multilayered cumulus cells **(Fig. 5)** adherent to zona pellucida, and homogeneous ooplasm. Oocytes can be degraded as:

A grade: Cumlulus oocyte complex (COC) with 4 or more layers of compact cumulus cells surrounding the zona pellucida and having an evenly granulated cytoplasm.

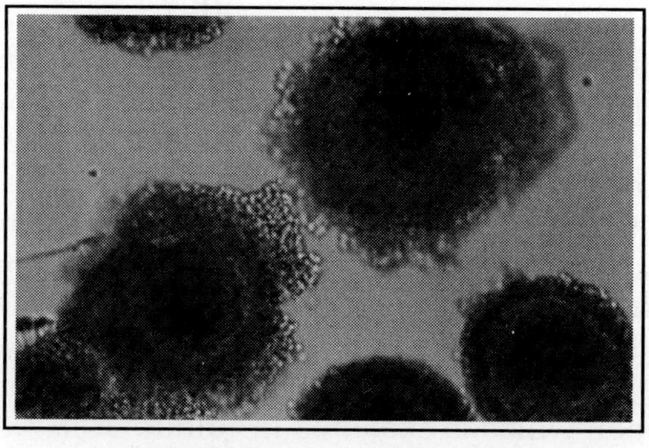

Fig. 5. Good quality oocytes

B grade: COC not having much compaction, with 1-3 layers of cumulus cells surrounding the zona pellucida and an evenly granulated cytoplasm.

C grade: Ocyte with fibrous (expanded) cumulus layers surrounding the zona pellucida.

D grade: Oocytes without cumulus cells and an irregular ooplasm. Generally A- and B-grade oocytes should be employed for *in vitro* embryo production (IVEP).

In terms of importance of cumulus cells, Cetica *et al.*, (1999) studied the nuclear stage of four grades of immature oocytes selected for maturation on the basis of cumulus-cell coverage; they found that 88% of the top-grade oocytes were at germinal vesicle (GV) stage, a figure significantly higher than in the other grades. It is also recognized that various components of follicular fluid, including progesterone and estrogen 17-β, produced by the cells of follicle, may play a pivotal role in the developmental competence of the oocytes. Lechniak *et al.* (2002) have confirmed that smaller oocytes tended to follow an abnormal path of meiotic maturation and result in disturbances in maturation process. As such the ability of oocytes to resume and complete meiosis in vitro is known to be related to its size and ploidy; the oocytes smaller in diameter than 110 μm and are still transcriptionally active and have a reduced ability to resume meiosis. Hence, use of smaller or under-developed oocytes should be avoided to obtain increased number of transferable quality embryos.

In vitro maturation

The fully grown mammalian oocytes before their rupture and ovulation, resume meiosis, progressing from prophase of first meiotic division to metaphase-II (M-II). This process involves complex changes in the transcriptome and proteome of the oocytes (Siemer *et al.*, 2009), promotion of cytoplasmic maturation (Ye *et al.*, 2009) is important in transforming the primary oocyte into a mature secondary oocytes. In the live animal, the oocyte, which has been arrested in the dicytate stage of the first meiotic division, resumes meiosis in response to the signals associated with the preovulatory luteinizing hormone (LH) surge (Gordon, 2003). *In vitro*, the oocyte maturation can be initiated by removal of a competent oocyte from its follicular environment; it is this phenomenon which is exploited in culture or *in vitro* maturation (IVM) system.

The aspirated buffalo oocytes are transferred into the droplets of *in vitro* maturation medium (TCM-199, FBS, antibiotics, buffalo follicular fluid, porcine FSH), overlaid with mineral oil. The buffalo oocytes are incubated in 5% CO_2 in humidified (90-95% relative humidity) air at 38.5°C ± 0.5°C, in CO_2 incubators.

After 24h of culture, the degree of expansion of cumulus cells of COC is examined and the cells exhibiting a uniform expansion **(Fig. 6)** constitute the preliminary criterion for ascertaining the oocyte maturation. Certain cumulus cells in the IVM medium droplets may adhere to the surface and form monolayer. These cells are allowed to form a cell bed which in combination with oviductal epithelial cell clumps is used in the embryo culture.

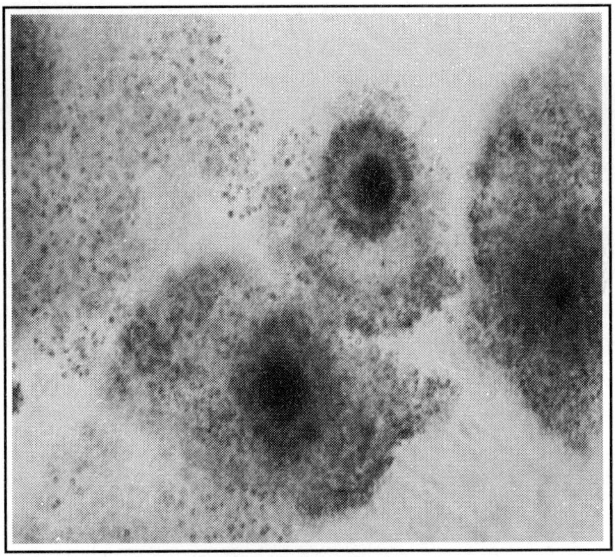

Fig. 6. COC exhibiting expansion

Sperm capacitation and preparation for IVF

Sperm while produced in the bull's testes are immobile, and do not actually move in the epididymis, their storage site. They become mobile only after ejaculation or when diluted in a suitable medium. The freshly ejaculated bull's sperm are incapable of fertilizing the oocytes until they have undergone a cascade of biochemical changes resulting in the state of 'capacitation'. In this process, the mammalian spermatozoa undergo several signaling and biochemical transformations.

Sperm processing still remains a critical factor in the process of assisted reproduction. In the early 1950s, Chang (1951) and Austin (1951) demonstrated independently that sperm had to be in the female reproductive tract for a finite period before acquiring fertilizing capacity. This phenomenon is known as "sperm capacitation". What made this finding a necessary step for the consequent development of IVF was the understanding that certain factors in the female reproductive tract were needed for the sperm to become fertile. A logical follow-up of the discovery of sperm capacitation occurred some years later when Chang (1959) demonstrated mammalian IVF conclusively by showing that eggs from a black rabbit were fertilized *in vitro* by capacitated sperm from a black male, and transferred to a white female, resulted in the birth of a litter of black offspring. In fact, the current advances in the human as well as livestock assisted reproduction can be attributed to the discovery of sperm capacitation which made the fertilization feasible *in vitro*, and revolutionized this very important field of the reproduction.

The capacitated sperm become capable of binding to the egg zona pellucida, where it undergoes the acrosome reaction (AR), a well-controlled process enabling it to penetrate and fertilize the egg. A very early event in sperm capacitation is the activation of sperm motility when they come in contact with high HCO_3^- and Ca^{2+} concentrations in the seminal fluid. *In vivo*, the female genital tract plays a vital and far complex role in controlling the sperm cpacitation and function, particularly in modulating motility and fertilization. Further, the process needs an active and specific coordination between the female tract, ovulation and the spermatozoa, and it is physiologically not complete until the sperm reach the oocyte. Phosphorylation puts the sperm in mood. Protein tyrosine phosphorylation is a key event accompanying sperm capacitation. The results show that caprine sperm displays an enhancement of phosphorylation in the tyrosine residues of specific proteins under *in vitro* capacitation conditions (Chatterjee *et al.*, 2009). The work by Morgan *et al.*, (2008) takes a chemical-genetic switch approach to understand the temporal action of an enzyme, protein kinase A (PKA) in sperm capacitation. A recent report (Pons-Rejarji *et al.*, 2009) presents the first global analysis of major elements individually described during bovine sperm capacitation and acrosome reaction signaling pathways. Another recent study on buffalo sperm capacitation has concluded the changes in buffalo spermatozoa in relation to seminal cholesterol, membrane fluidity and intracellular calcium levels (Kadirvel *et al.*, 2009). However, the molecular mechanisms underlying capacitation still remain elusive.

In vitro induction of capacitation

The capacitation can be mimicked *in vitro* by using appropriate capacitation media. Many reports are available on induction of capacitation in bovine sperm, using either chemical agents (high ionic strength in the medium, elevated pH, caffeine, heparin etc.) or exposing the sperm to oviductal cell culture. Heparin in the culture medium acts as inducer of capacitation *in vitro*. Heparin treatment is considered as the most important method for *in vitro* capacitation of bull sperm; the heparin is also believed to be secreted in the oviduct of female at the time of fertilization. The exact cellular and molecular mechanism of the bull capacitation is still a challenging area and needs further investigations. The quality and fertilization potential of the spermatozoa varies with batches of sperm collected from same or different bulls.

Use of fresh *vs.* frozen semen

The frozen semen is more desirable to avoid the variations in the batches. A number of sperm doses can be prepared from a single collection and cryopreserved for subsequent applications. It is recommended that in a single set of experiment, the semen should be from the same batch. This ensures homogeneity in the experimental data.

If frozen semen is used, the semen straws are frozen-thawed and washed twice with washing BO medium (BO medium containing 10 μg/ml heparin, 137.0 μg/ml sodium pyruvate and 1.942 mg/ml caffeine sodium benzoate). The sperm pellet is re-suspended in around 0.5 ml of the washing BO medium.

To exclusively select the motile sperm, a method called swim up **(Fig. 7)** is used which ensures higher IVF rate. The sperm are washed twice with BO medium and transferred to droplets containing 15-20 oocytes/droplet covered with sterile mineral oil, and then incubated in a CO_2 incubator (5% CO_2 in air) for 18 h at 38.5°C. After incubation with sperm, the oocytes are separated from the sperm droplets and washed with modified *in vitro* culture medium supplemented with FBS and fatty acid-free BSA. The medium needs to be replaced with

50% of fresh culture medium every 48 h. Generally, on day 8, the cattle or buffalo embryos reach hatching blastocyst stages.

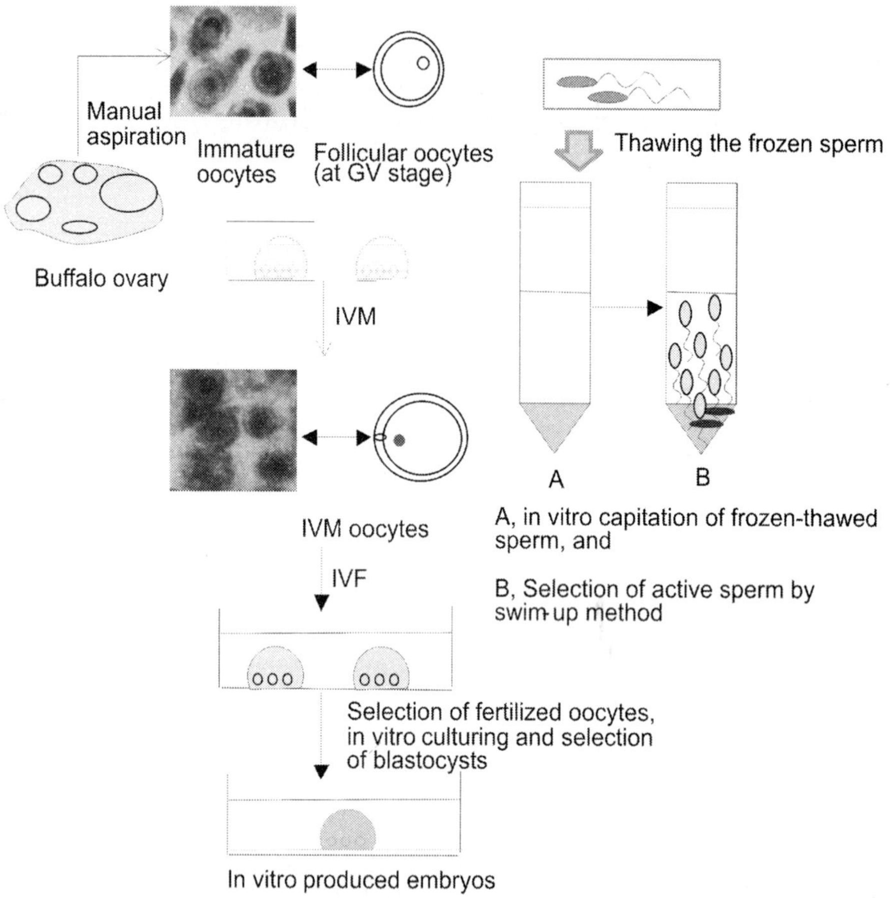

Fig. 7. A generalized plan for *in vitro* production of bubaline embryos after *in vitro* maturation (IVM), fertilization (IVF) and culture (IVC)

In vitro fertilization

In vitro matured oocytes are fertilized by the *in vitro* capacitated sperm. Various laboratories use their own standardized protocols or the standard protocols. In buffaloes the oocytes matured for 22-24 h are considered to be suitable for fertilization. For detailed information on IVF in buffaloes the readers are suggested to refer to publications including Madan *et al.* (1994) and Chauhan *et al.* (1998).

Intracytoplasmic sperm injection (ICSI)

ICSI is a relatively new ART that involves injection of a mature spermatozoon into a metaphase II oocyte, typically performed in cases of male factor infertility. As such, this technique bypasses multiple steps of the natural fertilization process by introducing an apparently intact spermatozoon into the ooplasm. The impact of the microinjection technique on fertilization and postfertilization events should be definitely established to determine the immediate safety of ART as well as any possible long-term consequences, including embryonic anomalies that result in developmental arrest or are carried over to the offspring.

Evaluation and preservation of embryos

Different methods are used for evaluation of *in vitro* produced embryos. These include the visual examination of embryos and counting the blastomeres **(Fig. 8)**. The blastomeres are first stained with Hoechst-33342, and the cells are fixed on the glass slides and visualized under a fluorescent inverted microscope.

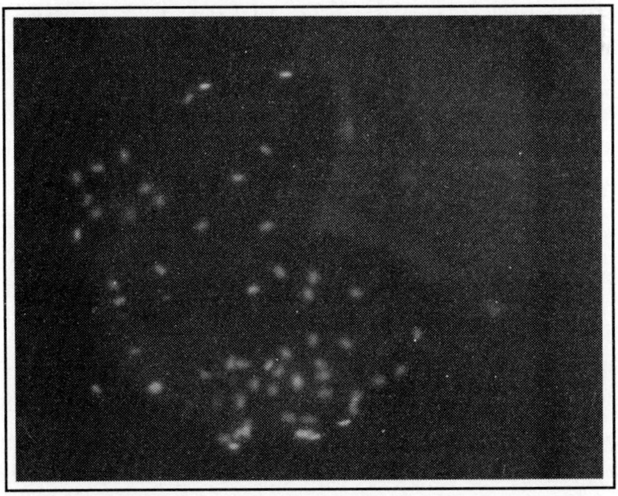

Fig. 8 Counting balstomeres using Hoechst 33342 staining

As a part of normal ET program, there is an obvious need for short-term preservation of the embryos recovered from the live animals or produced *in vitro*, until these embryos are transferred to the recipient or stored by cryopreservation. For practical reasons, the ability to store embryos at temperatures above freezing for relatively shorter periods

(24h-48h) may occasionally be preferable to freezing or vitrification. It is believed that during cryopreservation, the zona pellucida of the embryos may damage, making them more susceptible to pathogens and microbial infections.

Estrus synchronization and embryo transfer

Programmed breeding systems are beneficial to the management of reproduction of dairy cows because it is a method to schedule and control the insemination program of lactating dairy cows. The modal length of the estrous cycle is 21 days for mature cows and 20 days for heifers. Synchronization of estrus refers to the process of scheduling a group of recipient females to come into estrus at approximately the same time as donor animals which are being superovulated with exogenous hormones. Estrous synchronization gives the dairy and beef producers the opportunity to capture the economic benefits of artificial insemination.

Historically, estrus synchronization has been promoted as a labor saving tool for those who want to capitalize on the superior genetics available through the use of AI. The goal in controlling the bovine or bubaline estrus is efficient utilization of embryos produced by IVF or nuclear transfer. The proper timing or synchronization of estrus between the donor and recipient is important to obtain a pregnancy from embryo transfer. For this, it is a must that uterine environment must be at nearly similar stages for the process of conception. Protocols of estrus synchronization in buffaloes vary slightly among different laboratories. The most important aspect of estrus in buffaloes is lack of an accurate prediction and detection of estrus. Microscopic examination of vaginal mucus or help of teaser bulls is advisable for detection of estrus in buffaloes. Hence, estrus-synchronization programs should aim to synchronize follicular wave development as well as the onset of estrus and ovulation. The Ovsynch protocol (GnRH-7 d-PGF2α-2 d-GnRH- 16 h-TAI) is an effective planned breeding program that allows insemination of cows artificially without estrus detection.

Opportunities and challenges

Because the cost of ET is high, use is almost confined to the trade in high value pedigree animals. For all IVF technologies, achieving economically viable success rates is still a problem. Wastage of embryos during pregnancy is also increased, with IVF embryos achieving about

10% lower success rates than conventional embryos, where pregnancy rates of about 50% are regularly achieved. Freezing of embryos causes a further drop. Research is focused on minimizing such losses, as well as on improving the efficiency of oocyte recovery, IVM, IVF and embryo survival. Despite these difficulties, IVF laboratories are functioning in a number of countries, and research in recent years has concentrated on a new package of technologies. The buffaloes exhibit certain inherent reproductive problems which should be resolved for successfully exploiting the potential of the technology.

Conclusion

It is evident that healthy herds are more efficient. However, challenges in improving the success of assisted reproduction, broadening the availability and applicability of the technologies, and minimizing complications and risks continue to confront the field. The science of genetics, including epigenetics and affiliated branches, is undergoing an explosion in the development of new reproductive technologies and knowledge. These advances can and will improve and alter the current practice of assisted reproductive technology. The main objective of embryo transfer is to obtain more progeny per year from a single female of superior genotype. A country like India has a majority of animals with low productivity, and superior females are limited in number and highly priced. The assisted reproduction can fill the gap between supply and demand of elite livestock embryos for filed applications.

Table 1. Applications of embryo production and transfer technology

1. Achieving surprisingly rapid rates of multiplication of animals of the selected genetic superiority.
2. By combining the embryo splitting with superovulation, the rate of multiplication can be increased.
3. Increasing the number of calves from a given donor animal in any one year and enabling a varied number of sires to be used, possibly on that same flush or over a number of repeated flushes
4. Utilizing the ovaries and oocytes of the slaughtered females as a way of preserving genetic resources by *in vitro* embryo production.
5. The males produced through ET out of superior donors may be used as elite sires on AI network of livestock
6. Fulfilling the increasing demand for elite sires and dams for embryo production

Table 2. Advantages of OPU and IVEP technology in buffaloes

1. The technique can be used for recovery of oocytes from nearly all animals including infertile animals without the need of exogenous hormonal treatment for stimulation of donors
2. Aspiration of follicles can be done twice a week over a long period
3. In a period of one flushing, 10-12 OPU sessions can be applied which may result in about 14 transferrable-quality embryos and subsequently 6 pregnancies (Mishra, 2003).
4. Pre-pubertal donors at an early age of 11-12 months, and pregnant donors up to 3 months of pregnancy can be used as oocyte-donors
5. The technique enables exploiting the germplasm of the females that are incapable of breeding due to pathological problems or reproductive failures
6. Superior genotypes that are unfit to carry the fetus for full-term can be used as donors of the oocytes or embryos.

References

Alberio R, Palma GA, Stojkovik M, Brem G and Wolf E. **1998**. Effects of r-hCG and r-bGH plus r-hFSH in maturation medium on the developmental capacity of bovine oocytes. **In**: *Proc. 14th Meeting European embryo Transfer Association.* Vanice, pp.122.

Austin CR. **1951**. Observations on the penetration of the sperm in the mammalian egg. *Aust. J. Sci. Res.* **4**, 581-596.

Cetica PD, Pintos LN, Dalvit GC Beconi MT. **1999**. Effect of lactate dehydrogenase activity and isoenzyme localization in bovine oocytes and utilization of oxidative substrates on in vitro maturation. *Theriogenology.* **51**, 541-550.

Chang MC. **1951**. Fertilizing capacity of spermatozoa deposited into the fallopian tubes. *Nature.* **168**, 697-698.

Chang MC. **1959**. Fertilization of rabbit ova in vitro. *Nature.* **184**, 466-467.

Chatterjee M, Nandi P, Ghosh S and Sen PC. **2009**. Regulation of tyrosine kinase activity in goat sperm. *Mol. Cell Biochem.* (in press).

Chauhan MS, Singla SK, Palta P, Manik RS, and Madan ML. **1998**. *In vitro* maturation fertilization and subsequent development of buffalo (*Bubalus bubalis*) embryos: effects of oocyte quality and type of serum. *Reprod. Fertil. Dev.* **10**, 173–177.

Gordon I. **2003**. Laboratory Production of Cattle Embryos. Gordon, I (Ed.), 2nd Edition. CABI Publishing, CAB International Wallingford, Oxon OX10 8DE, UK.

Nagai T. **2001**. The improvement of in vitro maturation systems for bovine and porcine oocytes. *Theriogenology*. **55**, 1291-1301.

Jain GC, Das GK, Solanki VS and Tripathi VN. **1995**. Comparative efficiency of three different collection techniques for oocyte retrieval in buffaloes. *Theriogenology*. **43**, 240.

Kadirvel G, Kumar S, Kumaresan A and Kathivaran P. **2009**. Capacitation status of fresh and frozen-thawed spermatozoa in relation to cholesterol level membrane fluidity and intracellular calcium. *Anim. Reprod. Sci.* (in press)

Kumar A, Solanki VS, Jindal SK, Tripahthi VN and Jain GC. **1997**. Oocyte retrieval and histological studies of follicular population in buffalo ovaries. *Anim. Reprod. Sci.* **47**, 189-195.

Lechniak D, Kaczmarek D, Stanislawski, D and Adamowicz T. **2002**. The ploidy of in vitro matured oocytes is related the diameter. *Theriogenology*. **57**, 1303-1308.

Madan ML, Chauhan MS, Singla SK and Manik RS. **1994**. Pregnancies established from water buffalo (*Bubalus bubalis*) blastocysts derived from *in vitro* matured, *in vitro* fertilized oocytes and co-cultured with cumulus and oviductal cells. *Theriogenology*. **42**, 591–600.

Mishra AK. **2003**. Ovum pick-up from infertile farm animals. **In**: *Ovum Pick-up Technology for in vitro Production of Embryos from Farm Animals*. Chauhan MS, RK. Mahapatra, Majumdar AC, Sharma GT (Eds). Published by CAS-Veterinary Physiology Division of Physiology and Climatology, IVRI Izatnagar (India), pp. 34-40.

Morgan DJ, Weisenhaus M, Shum S, Su T, Zheng R, Zhang C, Shokat KM, Hille B, Babcock DF and McKnight GS. **2008**. Tissue-specific PKA inhibition using a chemical genetic approach and its application to studies on sperm capacitation. *Proc. Natl. Acad. Sci. USA*. **105**, 20470-20445.

Nicholas FW. **1979**. The genetic implications of multiple ovulation and embryo transfer in small dairy herds. **In**: *Proc. 30th Annu. Meet. European Assoc. Anim. Prod. (Harrogate)*, CG1.11.y

Pons-Rejarji H, Bailey JL, and Leclerc P. **2009**. Modulation of bovine sperm-signaling pathways: correlation between intracellular parameters and sperm capacitation and acrosome exocytosis. *Reprod. Fertil. Dev.* **21**, 511-524.

Siemer C, Smiljakovic T, Bhojwani M, Leiding C, Kanitz W, Kubelka M and Tomek W. **2009**. Analysis of mRNA associated factors during bovine oocyte maturation and early embryonic development. *Mol. Reprod. Dev.* **76**, 1208-1219.

Singla SK and Madan ML. **1990**. Response of superovulation in buffaloes (*Bubalus bubalis*) with Super-OV and FSH-P. *Theriogenology*. **33**, 327.

Sirard MA and Blondin P. **1996**. Oocyte maturation and IVF in cattle. *Anim. Reprod. Sci.* **42**, 417-426.

Smith AK. **2009**. Embryo transfer- opportunities for vets and scientists. *BCVA. Vol. 17* Part 1.

Ye Y, Kawamura K, Sasaki M, Kawamura N, Groenen P, Sollewijn Gelpke, MD, Kumagai J, Fukuda J and Tanaka T. **2009**. Leptin and ObRa/MEK signaling in mouse oocyte maturation and preimplantation embryo development. *Reprod. Biomed. Online.* **19**, 181-190.

—□□□—

5

Cryobanking

"Cells and tissues preserved under ultralow temperature can endure storage for centuries with almost no change in functionality or genetic information. Among the reproductive cells, oocytes present significant hurdles in achieving successful cryopreservation, primarily due to their sensitive microtubule structure. Novel methods of cryopreservation and their applications to preserve somatic cells, gametes, and embryos offer enormous opportunities in preserving and global dissemination and conservation of valuable livestock germplasm"

Introduction

Availability and preservation of raw biological materials are the major challenges in the widespread use of ARTs in livestock. To deal with the problem of uncertain supply and demand, the researchers often have to store gametes and embryos over extended period of time. For example, many AI procedures rely on availability of cryopreserved sperm. Similarly, cryopreserved oocytes and embryos are being increasingly used in assisted reproduction procedures. With the invention of advanced technologies like somatic cell nuclear transfer, animal transgenesis, and induced pluripotencey in normal somatic cells, the cryobanking has become significantly important. In addition, the cryopreservation serves as common way to germplasm conservation which involves the preservation of gametes and embryos, and reproductive tissues for use in artificial reproduction. The cryopreserved oocytes can be used for production of embryos through IVF, nuclear transfer and genetic manipulation.

Cryopreservation of mammalian embryos was initially shown to be feasible when Whittingham *et al.* (1972) and Festing (1972) obtained live mice pups after the transfer of frozen-thawed embryos that had been cryopreserved using either glycerol or dimethyl sulphoxide (DMSO). Since then the cryobiology has emerged as an important research area in biomedical and reproductive biological sciences. Owing to the immense practical importance, cryopreservation of sperm, oocytes and embryos has been extensively investigated, and still remains a fertile area for cryobiologists. Cryopreservation of immature oocytes leads to a new direction of egg banking in future. Similarly, cryopreservation of embryos, somatic and gonadal tissues has become increasingly important in biomedicine, reproductive biology and livestock conservation (Song *et al.*, 2009).

Principles of cryopreservation

It is evident that cellular metabolic activities decrease with drop in temperature and practically no thermally-driven chemical reactions occur at cryogenic temperatures. As such, the cells frozen and stored at cryogenic temperature such as liquid N_2 (-196°C) can endure storage for centuries with almost no change in functionality or genetic information (Mazur, 1984). During the cryopreservation, the cells or cellular systems exposed to sub-zero temperature are subjected to profound changes in thermal, chemical and physical properties in their surrounding milieu. As the ambient temperature drops from +37°C to -196°C, the concentration of solutes both inside as well as outside the cells increases dramatically. The cells can lose up to 95% of their water. Ice formation in the extracellular medium mechanically deforms and injures the cells, and intracellular ice formation causes the lethal injuries to the cells (Karlsson and Toner, 1996). Equally harsh is the process of rewarming the cells to physiological temperatures, during which the cells may encounter the hazardous osmotic shocks and the formation of large intracellular ice crystals by recrystallization (Mazur, 1984).

Cryoprotective agents

A cryoprotective agent (CPA) is a substance that is commonly added to freezing medium to prevent the biological material from freezing injuries i.e. damage due to ice formation. In most cases, little or no survival of cells is feasible without the application of CPA. Discovered accidently by Polge *et al.* (1949), the use of CPAs has since become ubiquitous in almost all cryopreservation protocols. Certain molecules are synthesized by the cells in multi-cellular organisms to protect them against extreme cold conditions. Under natural environmental conditions, Arctic and Antarctic insects, fish, amphibians and reptiles synthesize certain anti-freeze compounds and anti-freeze proteins in their body which act as physiological cryoprotectants, and minimize freezing damage to the host cells during cold winter. Among the insects, sugars and polyols act as anti-freeze compounds, whereas Arctic salamanders produce glycerol as cryoprotectant. The same principle is exploited in experimental and commercial cryobanking of the biological materials.

In experimental biology, the conventional CPAs are glycols (e. g. ethylene glycol and propylene glycol) and glycerol. Dimethyl sulfoxide (DMSO) is also regarded as a conventional cryoprotectant, and in combination with other compounds has been used for decades by the cryobiologists for cryopreservation of sperm and oocyte. CPAs can be broadly classified into two types permeating and non-permeating, depending on their ability to cross plasma membrane. The permeating type CPAs include methanol, glycerol, dimethylsulphoxide (DMSO), various glycols and high molecular weight alcohols. The commonly used non-permeating CPAs are sucrose, polyvinyl pyrrolidone, hydroxymethyl starch, dextran, serum and serum albumins), etc. The actual mechanisms by which CPAs protect the cells against cryo-injuries are still not clearly known. The permeating cryoprotectants are thought to reduce the amount of ice that forms in cell suspension during freezing. These CPAs reduce potentially harmful concentrations of electerolytes within the cell through their colligative properties (Ashwood-Smith, 1987). Other mechanisms of protection, such as stabilization of cell proteins (Arakawa *et al.*, 1990), and plasma membrane by direct interactions, have also been proposed (Anchordoguy *et al.*, 1991). The non-permeating cryoprotectants may exert their cryoprotective effects by their ability to reduce the concentrations of intracellular water, thus limiting the rate and extent of cellular crystallization and associated cell injury.

Methods of cryopreservation

The success of cryopreservation depends on procedures that minimize the formation of intra-cellular ice crystals when oocytes or embryos are held in solution. The reduced exposure time of the embryos to osmotic stress and toxic cryoprotectant leads to a high revival rate of the stored embryos. At the moment there are two major techniques for cryopreservation: slow-freezing and vitrification. The major difference between two is total avoidance of ice formation in vitrification.

Slow-freezing

A successful oocyte cryopreservation is of utmost importance where a limited number of oocytes are available to be manipulated or inseminated per cycle, to overcome legal and ethical issues related to embryo storage and for fertility preservation. Contrary to spermatozoa cryopreservation, oocyte cryopreservation currently has a low success rate. Embryo cryopreservation is used essentially with supernumerary embryos generated in ARTs. The conventional cryopreservation technique is a cooling technique involving different methods. It has certain technical limitations which make it time consuming and expensive as well. Hence, alternative strategies are being used.

Vitrification

Significant improvement of cryopreservation of oocytes has been achieved by both slow freezing and vitrification. However, cryopreservation using conventional slow-freezing methods has certain serious limitations like requirements of sophisticated and expansive equipments. To overcome the problems, an ultra-rapid cooling method known as vitrification or 'a glass-like solidification' is performed by suspending the biological materials in a solution containing a high concentration of cryoprotectants and then plunging them directly into liquid nitrogen (Rall and Fahy, 1985). The advantage of this technique is to prevent the formation of ice crystals within the biological material subjected to cryopreservation, though the toxic effects of high concentration of cryoprotectant media has been a concern. New vitrification techniques which attempt to accelerate the cooling rate by decreasing the cryosolution volume and concentration, may reduce the potential toxicity. In addition, a more rapid cooling rate results in reduced chilling injury (Vajta *et al.*, 1998). The success of vitrification can be attributed to the procedures that minimize the formation of intracellular ice crystals when an embryo/ oocyte and the surrounding vitrification solutions are "glassified". Also, the reduced exposure time of the embryos to osmotic stress and toxic cryoprotectant lead to high revival rate of the stored embryos or oocytes. The use of higher concentrations of CPAs increases the viscosity of the cell suspension, thus aiding vitrification at higher rates.

The minimum volume needed in vitrification facilitates increase of cooling and warming rates, reduction of concentration of cryoprotectants, and recovery of meiotic spindle. Thus vitrification appears to offer higher survival rate than the slow-freezing methods (Chen and Yang, 2009).

Since its invention, different devices are used as containers for pre-implantation embryo vitrification. Plastic straws (Otoi *et al.*, 1998) are widely used for vitrification. Open pulled straws (Vajta *et al.*, 1998) and the double straw system (Kuleshova and Shaw, 2000) were produced to improve sample cooling rate and to isolate samples from potential infectious agents accumulated in the storage vessels. Electron microscope grids (Martino *et al.*, 1996), Cryoloop (Lane *et al.*, 1999), Cryotop (Kuwayama *et al.*, 2005), nylon mesh (Matsumoto et al., 2001) and metal mesh (Fujino *et al.*, 2008) were recently adopted as an open devices to maximize the cooling rate. These devices had the limitation that they did not have a large holding capacity nor could they achieve a high embryo revival rate after vitrification. A recently reported vitrification spatula (VS) (Tsang and Chow, 2009) has overcome these limitations in that the VS is inexpensive, easy to assemble and can be readily made at home and has a large holding capacity to vitrify preimplantation mammalian embryos in a micro-drop employing ultra-rapid cooling in liquid nitrogen.

There are several technical challenges in developing successful cryopreservation for oocytes, most of which appear to lie at the sub-cellular level. Mature oocytes provide the better chance for embryo development, but these oocytes are arrested in metaphase II, the spindle apparatus is fully extended which may disassemble at lower temperature and, thus, may lead to chromosome disruption and aneuploidy (Roberts and Oktey, 2005; Koutalki *et al.*, 2006; Zenzes *et al.*, 2001). The technique is under improvement for increasing its applications. For instance, a modification of the conventional vitrification, called 'cover vitrification' with optimal concentration of cryoprotectants (Zhou *et al.*, 2009), and a novel "needle immersed vitrification" (NIV) (Wang *et al.*, 2008) with minimal volume of less concentrated CPAs are practically convenient compared to conventional vitrification, and has been expected to be efficient for cryopreservation of ovarian tissue with high efficiency and facility. Further, because of the necessity of programmable freezers required for slow-freezing and more commercial interest in the cryopreservation, the vitrification methods seem to have an edge over the slow-freeze methods.

Cryopreservation of somatic cells

Developments made in intracytoplasmic sperm injection (ICSI) and gamete isolation and maturation has generated interest in cryopreservation of gonadal tissues (i.e. testicular tissue and ovarian tissue) containing various developmental stages of male (spermatogonia, spermatids and spermatozoa) (de Watcher, 2005; Salzbrunn *et al.*, 1996) and female (primordial, secondary and tertiary) (Oktey and Sonmzer, 1994; Critser *et al.*, 2002; Roberts and Oktey, 2005) germ lines, as they appear to be promising avenues for various fertility treatments. Cryopreservation of somatic cells (Fig.1) has important application in somatic cell nuclear transfer technology, especially for solving the problem of supply of donor nuclei. Cumulus cells and fibroblasts have been vitrified from a number of animal species including experimental animals and farm animals like cattle (Gajda *et al.*, 2007) and buffaloes (Singh, 2006). This is an advancing area with novel applications. Frozen animal tissues without cryoprotectants have been thought to be inappropriate for use as nuclear donor for animal cloning. This concept appears to be inappropriate in view of the astonishing reports on production of viable cloned mice from the body frozen at -20°C for 16 years (Wakayama *et al.*, 2008), and recently, a bull cloned using the somatic cells retrieved from testicles that had been frozen without cryoprotectants at -80°C for ten years (Hoshino *et al.*, 2009). These studies clearly indicate that complete genome sets are maintained in animal organs even after long-term frozen-storage without CPA, and that somatic cells can be used as donor nuclei for resurrection of the extinct animals.

Fig. 1. Cryocontainers

Cryopreservation of buffalo oocytes and embryos

Oocytes present significant hurdles in achieving successful cryopreservation, primarily due to their sensitive microtubule structure. Efforts are continued to improve the cryopreservation of bubaline oocytes. Embryos developing to morulae and blastocysts obtained from slaughter-house derived oocytes have been successfully cryopreserved using slow-freezing or vitrification methods. However, IVF of frozen-thawed oocytes results in lower embryo production efficiency in comparison to IVF using fresh oocytes (without freezing) retrieved from abattoir-derived ovaries (Gautam *et al.*, 2008a, b).

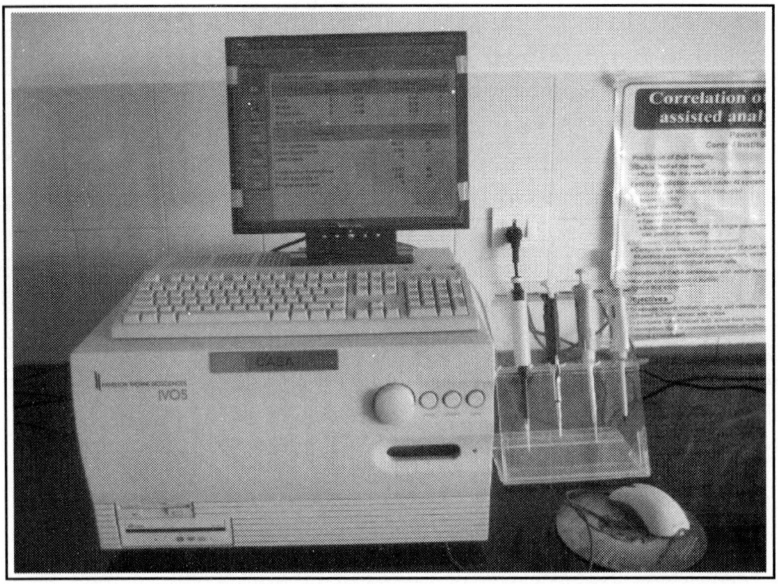

Fig. 2. Computer-assisted semen analyzer (CASA)

Conventional freezing and vitrification both are in use for cryopreservation of buffalo embryos **(Table-1).** Also, pregnancies from conventionally frozen embryos (Galli *et al.*1998, Zhang *et al.*, 2007) and from vitrified embryos (Negila *et al.* 2003), as well as calving from conventional freezing (Galli *et al.*, 1998, Zhang *et al.*, 2007), and vitrification (Hufana-Duran *et al.*, 2004) have already been established.

Blastocyst hatching rates following vitrification of embryos produced from the oocytes collected by OPU are significantly higher than the oocytes collected from slaughterhouse ovaries. The study showed that source of oocytes (OPU vs. abattoir ovaries) affects the *in vitro* embryo development and blastocyst hatching rates following vitrification of embryos in buffaloes (Manjunatha *et al.*, 2008c). These findings emphasize the feasibility of establishing oocyte banks for embryo production. Improving the viability of oocytes following cryopreservation, and using the frozen-thawed oocytes for producing cloned embryos in buffalo would be practically advantageous, as a continued supply of oocytes is needed in the process. Another study on quality analysis of buffalo cloned embryos derived from oocytes vitrified before or after their enucleation, indicates that vitrified-thawed oocytes could support development of cloned blastocycts, albeit with lower efficacy (in terms of number of blastomeres per embryo) compared to those derived from freshly retrieved oocytes (Muenthaisong *et al.*, 2007). Laowtammathron *et al.*, (2005) have reported that bovine cloned blastocysts, regardless of their hatching stage, were relatively resistant to vitrification by ultra-rapid cooling procedure when they were produced in presence of linoleic acid-albumin in an *in vitro* culture medium. The cloned buffalo embryos were more tolerant to vitrification than cloned cattle embryos. This report is encouraging as the surplus cloned buffalo embryos can be vitrified and used for transfer to recipients or used for further *in vitro* studies. A study on evaluating the developmental competence of *in vitro* matured buffalo oocytes after vitrification has revealed that oocytes subjected to vitrification do retain their developmental competence to grow to blastocyst stages *in vitro* (Attanasio *et al.*, 2009).

Opportunities and challenges

Over the past few decades, cryopreservation of reproductive cells and tissues has witnessed tremendous progress, thus offering a range of fertility treatment and fertility preservation options for improving livestock production. However, significant technical challenges still remain and attempts to address these problems in the coming years will constitute the bulk of research thrust in this area. The conventional method of slow-freezing requires expensive equipments **(Fig. 2)**, and at the same time certain cells are more sensitive to this method of cryopreservation. Vitrification is under refinement and used as alternative to conventional freezing. At the moment, the major limitation of vitrification is toxicity of the CPAs to biological materials used. Both techniques need optimization for their applications to preserve the biological materials for veterinary applications. Recent innovations on resurrection of the animals from the cells preserved without CPA suggest that it may be possible to exclude or minimize the use of chemical agents in the ultralow preservation of biological material while maintaining their viability. It is an emerging area with enormous potential, and needs a prioritized attention by the cryobiologists.

Conclusion

Cryopreservation of gonadal tissue and oocytes has brought about new and exciting research in reproductive medicine, as well as new cryopreservation techniques that are dramatically improving post-thaw viability and freezing damage. Embryo cryopreservation is an indispensable technique in mammalian reproductive programs and in animal facilities and in the genetically modified experiments. The techniques of cryopreservation are yet quite inefficient and need optimization for larger scale applications. In view of its simpler technical and equipment requirements the vitrification appears to be the method of choice in near future. It is expected that vitrification will become more successful by ensuring higher survival of the oocytes for applications in embryo production *in vitro*.

Table 1. Salient features of cryopreservation technology to bubaline reproduction

1. Global dissemination and preservation of valuable germplasm, embryos and somatic cells
2. Management of supernumerary embryos or embryos destined to be transferred under more appropriate conditions
3. Freezing of embryos serves as an alternative to limited number of recipients and preserving specific genotypes or mutants for the future.
4. Increasing the interval between time of collection and transfer or in vitro production and the transfer to estrus synchronized females.
5. Preserving the valuable female germplasm from the slaughtered females
6. Ensuring uninterrupted supply of oocytes for embryo production
7. Preserving the somatic cells for use as donor nuclei or production of induced pluripotent stem cells.

References

Anchordoguy TJ, Cecchini CA, Crowe JH and Crowe M. **1991**. Insights into cryoprotective mechanisms of dimethyl sulphoxide for phospholipid bilayers. *Cryobiology*. **28**, 467-473.

Arakawa T, Carpenter JE, Kita YA and Crowe JH. **1990**. The basis of toxicity of certain cryoprotectants. *Cryobiology*, **27**, 401-415.

Ashwood-Smith MJ. **1987**. Mechanism of cryoprotaction. **In**: *Temperature and Animal cells*. Bowler K. And Fuller BJ (Eds.). The Company of Biologists, Ltd. Cambridge, UK, 395-406.

Attanasio L, Boccia L, Vajta G, Kuwayama M, Campanile G, Zicarelli, L, Neglia G and Gasparrini B. **2009**. Cryotop vitrification of buffalo (*Bubalus bubalis*) *in vitro* matured oocytes: effects of cryoprotectant concentrations and warming procedures. *Reprod. Domest. Anim*. (in press).

Chen SU and Yang YS. **2009**. Slow-freezing or vitrification of oocytes: their effects on survival and meiotic spindles, and the time schedule for clinical practice. Review. *Taiwan J. Obstet. Gynacol*. **48**, 15-22.

Critser JK, Agca Y and Woods EJ. **2002**. Cryopreservation of immature and mature gametes . **In**: *Assisted Reproduction Technology. Accomplishments and New Horizons*. De Jonge CJ and Barrat CLR (Eds.). Cambridge University Press. UK. 144-166.

de Watcher MA. **2005**. Ethical aspects of cryobiology: responsible applications in biomedicine and clinical practice. *Cryobiology*, **48**, 205-213.

Festing M. **1972**. Mouse strain identification. *Nature*. **238**, 351-352.

Fujino Y, Kojima T, Nakamura Y, Kobayashi H, Kikuchi K and Funahashi H. **2008**. Metal mesh vitrification (MMV) method for cryopreservation of porcine embryos. *Theriogenology*. **70**, 809-817

Gajda B, Katska-Ksiazkiewicz L, Rynska B, Bochnek M and Smorag Z. **2007.** Survival of bovine fibroblasts and cumulus cells after vitrification. *CryoLetters*. **28**, 271-79.

Galli C, Duchi R, Crotti G and Lazarri G. **1998**. Embryo production by ovum pick up in water buffalo. *Theriogenology* **49**, 400.

Gautam SK, Verma V, Singh B, Palta P, Singla SK, Chauhan MS and Manik RS. **2008**a. Effect of slow-freezing on morphology and developmental competence of buffalo (*Bubalus bubalis*) immature oocytes. *Anim. Reprod. Sci*, **105**, 311-318.

Gautam SK, Verma V, Palta P, Chauhan MS and Manik RS. **2008**b. Effect of type of cryoprotectant on morphology and developmental competence of *in vitro* matured buffalo (*Bubalus bubalis*) oocytes subjected to slow freezing or vitrification. *Reprod. Fertil. Dev*. **20**, 490-496.

Hoshino Y, Hayashi, N, Taniguchi S, Kobayashi N, Sakai K, Otani T, Iritani A and Saeki K. **2009.** Resurrection of a bull by cloning from organs frozen without cryoprotectant in a freezer for a decade. *PLos ONE*. 4, e4142.

Hufana-Duran D, Pedro PB, Venturina HV, Hufana RD, Salazar AL, Duran PG and Cruz LC. **2004.** Post-warming hatching and birth of live calves following transfer of *in vitro*-derived vitrified water buffalo (*Bubalus bubalis*) embryos. *Theriogenology*. **61**, 1429–1439.

Karlsson JO, Toner M. **1996**. Long-term storage of tissues by cryopreservation: critical issues. Review. *Biomaterials*. **17**, 243-256.

Koutalki M, Schoepper B, Maroulis G, Diedrich K and Al-Hasani S. **2006.** Human oocyte cryopreservation: past present and future. *Reprod. Biomed. Online*. **13**, 427-436.

Kuleshova LL and Shaw JM. **2000**. A strategy for rapid cooling of mouse embryos within a double straw to eliminate the risk of contamination during storage in liquid nitrogen. *Hum. Reprod*. **15**, 2604-2609.

Kuwayama M, Vajta G, Ieda S and Kato O. **2005**. Comparison of open and closed methods for vitrification of human embryos and the elimination of potential contamination. *Reprod. Biomed. Online*. **11**, 608-614.

Lane M, Schoolcraft WB and Gardner DK. **1999**. Vitrification of mouse and human blastocysts using a novel cryoloop container-less technique. *Fertil. Steril.* **72**, 1073-1078.

Laowtammathron C, Lorthongpanich C, Ketudat-Cairns M., Hochi S, and Parnpai R. **2005.** Factors affecting cryosurvival of nuclear transferred bovine and swamp buffalo blastocysts: effects of hatching stage, linoleic acid–albumin in IVC medium and ficoll supplementation to vitrification solution. *Theriogenology*. **64**, 1185–1196.

Manjunatha BM, Gupta PS, Ravindra JP, Devraj M and Nandi S. **2008**. *In vitro* embryo development and blstocyst hatching rates following vitrification of river buffalo embryos produced from oocytes recovered from slaughterhouse ovaries or live animals by ovum pick up. *Anim. Reprod. Sci.* **104**, 419-426.

Martino A, Songsasen N and Leibo SP. **1996**. Development into blastocysts of bovine oocytes cryopreserved by ultra-rapid cooling. *Biol. Reprod.* **54**, 1059-1069.

Matsumoto H, Jiang JY, Tanaka T, Sasada H and Sato E. **2001**. Vitrification of large quantities of immature bovine oocytes using nylon mesh. *Cryobiology*. **42**, 139-144.

Mazur P. **1984**. Freezing of living cells: mechanisms and implications. *Am. J. Physiol.* **247**, C-125-C142.

Muenthaisong S, Laowtammathron C, Ketudat-Cairns M, Parnpai R, and Hochi S. **2007**. Quality analysis of buffalo blastocysts derived from oocytes vitrified before or after enucleation and reconstructed with somatic cell nuclei. *Theriogenology*. **67**, 893–900.

Neglia G, Gasparrini B, Caracciolo di Brienza V, Di Palo R, Campanile, G, Presicce GA and Zicarelli L. **2003**. First pregnancies established from vitrified blastocysts entirely produced in vitro in Mediterranean Italian buffalo cows. *Theriogenology*. **59**, 374.

Oktey K and Sonmezer M. **1994**. Ovarian tissue banking for cancer patients. *Hum. Reprod.* **19**, 477-480.

Otoi T, Yamamoto K, Koyama N, Tachikawa S and Suzuki T. **1998.** Cryopreservation of mature bovine oocytes by vitrification in straws. *Cryobiology*. **37**, 77-85.

Polge C, Smith AU and Parkes A. **1949**. Revival of spermatozoa after vitrification and dehydration at low temperatures. *Nature*. **164**, 166.

Rall WF and Fahy GM. **1985**. Ice-free cryopreservation of mouse embryos at -196 degree C by vitrification. *Nature*. **313**, 573-575.

Roberts JK and Oktey K. **2005**. Fertility preservation: a comprehensive approach to the young women with cancer. *J. Natl. Cancer Inst. Monogr*. **34**, 57-59.

Salzbrunn A, Benson DM, Holstein AF and Schulze W. **1996**. A new concept for the extraction of testicular spermatozoa as a tool for assisted fertilization (ICSI). *Hum. Reprod.* **117**, 752-755.

Singh B. **2006**. Studies on somatic cell nucleus transfer in buffaloes. Ph. D. Dissertation, submitted to National Dairy Research Institute (Deemed University), Karnal-132 001. India.

Song Y, Sharp R, Lu F and Hassan M. **2009**. The future potential of cryopreservation for assisted reproduction. *Cryobiology*. (in press).

Tsang WH and Chow KL. **2009**. Mouse embryo cryopreservation utilizing a novel high-capacity vitrification spatula. *Bio-Techniques*. **46**, 550-552.

Vajta G, Holm P, Kuwayama M, Booth PJ, Jacobsen H, Greve T and Callesen H. **1998**. Open pulled straw (OPS) vitrification: a new way to reduce cryoinjuries of bovine ova and embryos. *Mol. Reprod. Dev*. **51**, 53-58.

Wakayama S, Ohta H, Hikichi T, Mizutani E, Iwaki T, Kanagawa O. and Wakayama T. **2008.** Production of healthy cloned mice from the body frozen at -20 degrees C for 16 years. *Proc. Natl. Acad. Sci. USA*. **105**, 17318-17322.

Wang Y, Xiao Z, Li L, Fan W and Li S. **2008**. Novel needle immersed vitrification: a practical and convenient method with potential advantages in mouse and human ovarian tissue cryopreservation. *Hum. Reprod.* **23**, 2256-2265.

Whittingham DG, Leibo SP and Mazur P. **1972**. Survival of mouse embryos from -196°C to -269°C. *Science*. **178**, 411-412.

Zenzes MT, Bielecki R, Casper RF and Leibo SP. **2001**. Effects of chilling to 0 degrees C on the morphology of meiotic spindles in human metaphase II oocytes. *Fertil. Steril*. **75**, 769-777.

Zhang X, Liang X, Yang B, Chen M, Huang F, Pang C, Qin G, Senatore EM and Presicce GA. **2007**. Pregnancies established following transfer of *in vitro*-produced fresh and frozen-thawed buffalo (*Bubalus bubalis*) embryos. *Reprod. Fertil. Dev*. **19**, 276-276.

Zhou XH, Wu YJ, Shi J, Xia YX and Zheng SS. **2009**. Cryopreservation of human ovarian tissue: comparison of novel direct cover vitrification and conventional vitrification. *Cryobiology*. (in press).

—□□□—

6

Sex Pre-Selection

"If the sex of an embryo could be determined in advance, this could be an advantage to the end user, and hence adds to value of the product. With present technologies, it is possible to extract one cell from an early embryo, and with the use of a DNA probe to know whether it is male or female. Applications of sexed-sperm and pre-implantation genetic diagnosis allow dairy producers to select among their herd's potential dams and produce dairy replacement heifers from only the genetically superior females. As the dairy farmers have less uses of male calves, use of sexed-sperm to produce only females would make buffalo production more attractive. On the other hand, as male calves exhibit a higher growth rates and could be used as meat or draft animals"

Introduction

In normal mammalian male ejaculate, the half of the total sperm contains X-chromosome which upon fertilization results in female embryos, while the other half contains Y-chromosome resulting in a male embryo after fertilization. Sperm bearing X-chromosome contains around 3.0 to 7.55% more genomic DNA than the sperm bearing Y-chromosome depending on the species. Procedures are now available that can precisely identify and differentiate X- and Y-chromosome bearing sperm and separate them. The separated sperms can be used for *in vitro* production of embryos. Alternatively, the embryos produced *in vitro* or *in vivo* can also be sexed using cellular and molecular biological tools, and the selected embryos can be cryopreserved, disseminated and transferred into the surrogate mothers to produce a calf of desired sex. In several countries, embryo sexing is being applied at commercial level in companies and farms working with embryo transfer technology.

Pre-implantation genetic diagnosis

Prenatal diagnostic technologies like ultrasound scanning and amniocentesis have proved profitable when marketed as a method of sex selection. The ability to biopsy embryos and advances in the field of molecular genetics have allowed for pretransfer analysis of the genetic make-up of embryos created through *in vitro* fertilization. Pre-implantation genetic diagnosis (PGD) - a technically complex protocol is defined as the testing of pre-implantation stage embryos or oocytes for genetic defects. Originally, the technique had been developed for human couples whose potential offspring were at risk of severe

Mendelian disorders, structural chromosomal abnormalities or mitochondrial disorders (Ceraedts and Wert, 2009). The PGD developed in 1990s (Handyside *et al.*, 1990), as a part of progress in reproductive medicine, genetic and molecular biology is now used as a promising method to analyze embryos genetically before their transfer.

How is PGD performed?

Pre-implantation genetic diagnosis is an established procedure of embryo genetic diagnosis, and requires a close collaboration between obstetricians, fertility specialists, IVF laboratory and geneticists. Basically, the methodology involves biopsy from the embryo and its genetic analysis, though the techniques may vary for different modifications. With the advent of technologies taking cell(s) from the embryos and the advanced genetic analysis methods, it is now possible to select the healthy embryos for future applications. Four methods are described here.

Fluorescent *in situ* hybridization (FISH)

A procedure known as "*in situ* hybridization (ISH)" wherein chromosome-specific DNA probes are hybridized to cell nuclei is used in genetic diagnosis of the mammalian cells. Although accuracy of this approach is acceptable, it requires days rather than hours, and the embryos have to be frozen till diagnosis. A modified version of ISH, known as fluorescent *in situ* hybridization (FISH) has made the technique much faster. Standard IVF technique is used when procedure requires FISH analysis. ICSI can be an option when there is abnormal semen or PCR analysis is required. This is to reduce the risk of contamination with sperm DNA. A single cell is removed from developing morula or 8-cell stage embryo through an opening in zona pelucida. The cell is then analyzed for the presence of genetic disorders. The FISH allows specific chromosome to be identified by utilizing the chromosome-specific fluorescent probes. The labeled probes bind to chromosome and can be visualized under a fluorescent microscope. This technique is used for sex determination in sex-linked diseases and for numerical and structural chromosomal abnormalities e.g., aneuploidy screening.

Polymerase chain reaction (PCR)

Development of PCR and synthesis of gene-specific primers in the early 1990s opened up new possibilities for embryo-sexing. The PCR diagnosis can be suitably performed within a day or so of the test and allows the amplification of selected nucleotide sequence from the genomic or mitochondrial DNA and is used to detect mutation of a single gene. The technique has been used to sex the embryo in livestock species including cattle, buffalo, horse, sheep and mice, etc.

Since the choice between male or female development in mammals is controlled by the sex chromosomes, several approaches to sex determination involve co-amplification of a Y-chromososmal sequence, commonly conserved in mammals, and an autosomal sequence that acts as a control for the presence of DNA and appropriate conditions for amplification (Appa Rao *et al.*, 1994; Park *et al.*, 2001; Mara *et al.*, 2004). Embryo sexing by PCR requires a small amount of DNA. It is a fast, reliable and accurate technique which enables the use of embryos for transfer or preservation for subsequent applications.

Pre-implantation genetic haplotyping (PGH)

PGH is a procedure that can weed out genetically defective embryos before they have a chance to initiate a pregnancy. This is particularly required when there is doubt about passing an incurable genetic disease or disorder to the offspring. This "*technically-demanding and complex procedure*" was developed only recently (http://www.healthlibrary.com, accessed on Oct. 23, 2009). It uses DNA fingerprinting to identify chromosomes carrying affected genes. For PGH, the females are superovulated and embryos are produced by IVF. Blastomere is taken from the embryos when they are at 8-16 cell stage of development. The DNA of the blastomere is diagnosed for genetic abnormality using PCR and fingerprinting. If a genetic defect is found, then the embryo from which it was taken is destroyed. Currently, the technique is only available in a few human clinics worldwide, but may have veterinary applications in near future.

Comparative genomic hybridization (CGH)

The CGH is a new technique which allows removal of cells from developing embryos at a stage when embryo is 5-6 days old, and at 64-128 cell stage. Removing blastomere at this stage is comparatively less damaging to the embryo. CGH together with microarrays analysis can improve PGD effectively.

First or second polar body can be used to study maternal genetic contribution or defects.

PGD is recognized as an alternative to prenatal diagnosis for the detection of genetic disorders in the species at risk of transmitting genetic abnormalities to their offspring. In livestock species, the technique offers opportunity to screen and discard the embryos carrying any genetic abnormality. However, diagnosis from a single cell remains a technically challenging procedure and the risk of misdiagnosis cannot be eliminated.

Significance of sex or gender pre-selection in livestock production

The ability to pre-determine the sex of a calf has been a long-standing goal of the dairy and beef industries (Moore and Thatcher, 2006), and currently ranks as one of the most desirable reproductive technologies. Pre-selecting the sex of animals used for production agriculture is particularly valuable in instances where one sex is more valuable in the production chain. For example, in commercial dairy system with the exception of a few bull-calves produced from the very top dams, most of the male calves have little role in the herd. On the other hand, the value of heifer calves is at a premium. The opposite is true in commercial beef production program. Sperm-sorting would allow commercial meat producers to select semen which when used would produce the offspring of desired sex in majority of the animals. This technology would also benefit producers using crossbreeding systems that employ maternal (F-1 animals) and/or terminal cross lines because only a minimum number of females would be needed to produce crossbred replacements or slaughter animals depending on production goals.

Until 1990s, the only option was to sex the embryos and transfer the sexed-embryos to recipients, but the techniques are currently available for separating the X- and Y- bearing sperm and subsequently using them for IVF or ICSI. It was not until the flow-cytometric sorting system was improved greatly and methods for cryopreservation of sex-sorted sperms were developed, which assisted in commercializing the sexed semen technology worldwide. A new adaptation of a system for separating X- and Y-chromosomes bearing spermatozoa has been developed by researchers in Germplasm and Gamete Physiology Laboratory at the US Department of Agriculture in Beltsville, Maryland, USA. The new technique, developed by Lawrence Johnson and colleagues builds on the Beltsville Sperm-sexing Technology which was patented in 1989. These methods rely on difference between DNA contents in X- and Y-chromosomes. The original technique could be used in rabbits, pigs, and cattle, but had the limitation that only 1.5-2 million sperm could be sorted per day and accuracy varied within 75-90%. The technique has been modified and its effectiveness has already been tested in sheep, laboratory animals and other species. In addition, commercially available post-thaw bovine sexing kit (e.g. HeiferPlus, HP) claims to alter the sex ratio in favor of female calves following artificial insemination (Curry *et al.*, 2009).

Availability of the sexed-sperm enables the livestock owners to predetermine the sex of their livestock, and thus, offers remarkable management flexibility. The widespread commercial application of sexed-semen is expected within next decade because of improvements in sperm-sorting and fertilization capacities of the sexed semen (Cran, 2007; De Vries *et al.* 2008). The PGD, therefore, allows the continued use of valuable germplasm even from carrier animals by screening embryos before transfer and discarding the few affected embryos.

Sperm sorting: Principle and methodology

As stated above, the DNA contents in X- and Y-chromosomes vary slightly and the difference can be detected and measured. Flow cytometry (measurement of cells as they flow by a detector) **(Fig. 1)** has been available for analysis and sorting a variety of cell types in fluid suspension since the late 1970s. The sperm separation is based on staining the chromosomes and then separating them into different types. Sperm cells are exposed to DNA binding stain for about 45 minutes. Following staining, sperm cells enter the flowcytometer where a vibrating peizoelectric crystal separates each cell into a single droplet. The droplets then pass through the laser beam in single file. The cells fluoresce at different intensities when exposed to a specific wavelength of blue light. Cells with greater fluorescence would be X-chromosome-bearing because more stain would be incorporated into a cell with greater quantities of DNA available for binding. After chromosome content is determined, either a negative or positive charge is applied to each sperm cell. Sorting can then take place as the sperm cells are attracted to either the positive or negative plate depending on which charge was previously applied. A recent report by Blondin *et al.* (2009) has described tools that can monitor semen parameters to optimize IVF conditions for obtaining maximal blastocyst rate. It has been shown that sexed-semen that appeared more capacitated seemed to require less heparin in IVF than sexed-semen that appeared less capacitated to produce a maximal percentage of blastocyst. Further, flow cytometric sorting in bovine eliminates spermatozoa that possess compromised DNA and, therefore, explains that the reduced fertility seen *in vitro* is not due to an increased percentage of spermatozoa with compromised DNA (Blondin *et al.*, 2009).

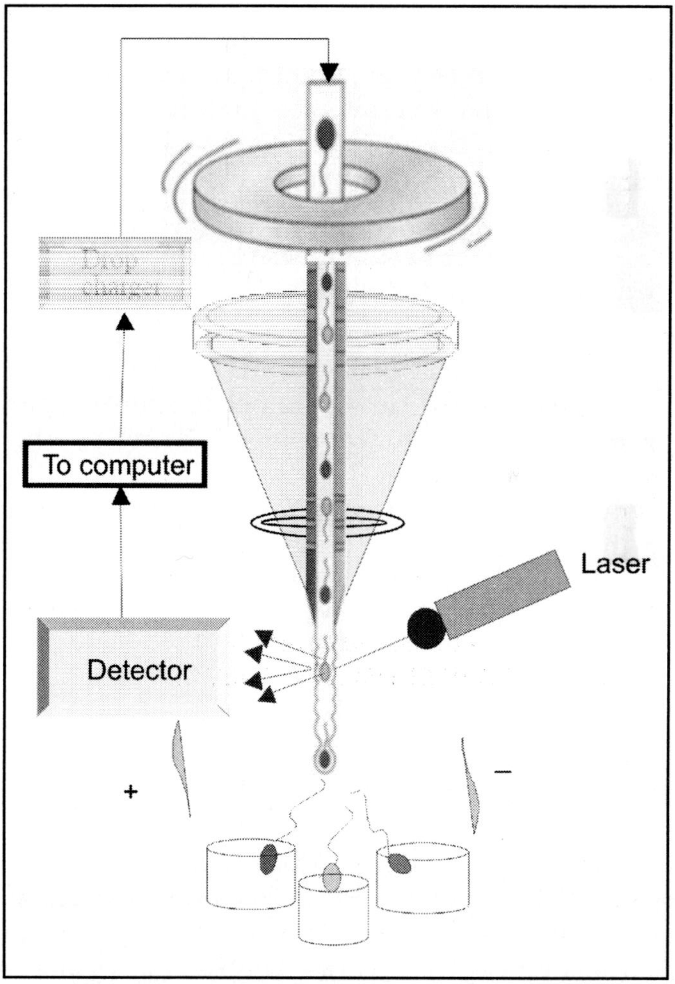

Fig. 1. Flow-cytometer method for Sperm-sorting

Cytomation Inc., a Fort Collins, CO, USA, Biotechnology Company, had acquired the rights of the high speed flow-cytometer/sorter that had been developed at Lawrence Livermore National Laboratory by Van den Engh and Stokdijk (Van den Engh and Stokdijk, 1989). This system was commercialized at Cytomation as the MoFloTM cytometer. The standard MoFlo1 was modified as a sperm sorter by using a beveled injection needle similar to what had been originally developed by Fulwyler (1977) and used by Pinkel *et al.* (1982a, b) to orient sperm. This instrument, MoFlo SXTM can precisely measure sperm DNA content from the flat surface of the sperm head.

No method, other than flow-sorting, has been shown to repeatedly produce viable sexed-sperm on a level necessary for commercialization. The commercial success of flow cytometric separation of X- and Y-sperm has stimulated renewed interests in developing other, perhaps more efficient methods, to pre-select the sex of mammalian offspring. It is probable that simpler, better systems will be developed for sexing sperm.

Current state-of-the art in water buffaloes

Sperm sexing technology in buffalo is in initial stages of validation and only a few reports are available at present. Murrah and Nili-Ravi buffaloes were selected for flow cytometric sperm separation and identifying the X- and Y-chromosomes bearing sperm based on difference in their DNA content (Lu *et al.*, 2006). The results prompted use of flowcytometric sorting of X- and Y-sperm in buffaloes. Continuing the studies further, Lu *et al.* (2007) reported that flow-cytometric sperm sorting accuracy for X- and Y-sperm was 94% and 89%, respectively. The sorted sperms were then used for IVF to produce sex-preselected embryos. Transplantation of the presumed X-embryos derived from IVF (using sexed sperm) into a recipient resulted in the birth of female twins, indicating the potential of the technology in buffaloes in terms of feasibility of sperm-sorting by flowcytometer, *in vitro* production of sex-preselected embryos and offspring_(Lu *et al.*, 2007). On exploring the use of sexed-sperm and OPU-derived oocytes in an IVEP system to produce sex-preselected buffalo embryos, Liang *et al.* (2008) reported that IVF of OPU-derived oocytes with sexed sperm had similar developmental competence, as those from the oocytes derived from abattoir-derived ovaries. Higher embryo cleavage (57.6% *vs.* 50.4%) but low blastocyst development (16.0% *vs.* 23.9%) rates were obtained in OPU-derived vs. slaughterhouse-derived oocytes. From the data, it was concluded that using frozen-thawed sexed *vs.* unsexed semen, the cleavage rates and blastocyst developmental abilities were unaffected in buffalo (Liang *et al.*, 2008). The study thus, provides a proof of concept for further research and wider field applications of these technologies in buffalo. Garner and Seidel (2008) emphasize that molecular cellular methods (e.g. flow-cytometric sorting) permit not only the faster differentiation of X- and Y-sperm, but also allows obtaining relatively pure X- and Y- sperm populations without obvious cellular damages.

The ability to sex the semen has a large potential for commercialization; thus research to develop and refine sperm-sexing technology is being carried out in the private sector as well. Innovative approaches by a diverse group of researchers along with advances in computer science, biophysics, cell biology and instrumentation etc., provide optimism and authentic basis for commercializing the application of sexed-semen in livestock reproduction (Garner and Seidel, 2008). Concerns regarding the use of sexed-semen technology include the apparent lower fertility of sorted sperm, the lower survival of sorted sperm after cryopreservation and the reduced number of sperms that could be separated in a specified time period (Lonergan, 2007). The use of sexed semen following superovulation in elite females will benefit the buffalo breeding managers in obtaining the calves of desired sex for milk or meat production.

Currently, the technology is in use to a limited extent but the use is likely to expand in future for a want of male or female buffaloes, if the biological causes of impaired embryo development and compromised conception rates of transferred embryos are investigated. Buffalo sperm-sexing in India is in the process of beginning. Animal Biotechnology Centre at National Dairy Research Institute, Karnal (India) has started the sperm sexing in water buffaloes. The use of sex-sorted buffalo sperm in IVF needs to be optimized because use of this approach to sex pre-selection is likely to increase substantially.

Additionally, modern molecular tools are proving to be extremely reliable and useful in sexing the embryo prior to their transfer to the recipients (Fu *et al.*, 2007). For instance, duplex PCR assay (Prashant *et al.*, 2008) and loop-mediated isothermal amplification (LAMP), a novel DNA amplification method that amplifies a target sequence specifically under isothermal conditions, have been developed for rapid sexing of buffalo embryos through identification of Y-chromosome specific sequences (Hirayama *et al.*, 2006; Zhang *et al.*, 2009). The LAMP has been reported to be reliable, quick (taking only 45 min.) and applicable as the total time for embryo sexing including DNA extraction was about 1h, and no thermal cycler or electrophoretic analysis was needed in the process (Hirayama *et al.*, 2006).

Fig. 2. Flow cytometer

Opportunities and challenges

Use of sexed sperm could help farmers get the animals of the desired sex. Despite enormous impact on livestock production and management, the sperm sexing technology has certain drawbacks which limit its larger application. First, the sperm separation efficiency is low. Second, the technology is expensive and requires high cost equipment, the fluorescent activated cell sorter (FACS) **(Fig. 1&2)**. The low sorting efficiency is a major problem, especially in species that require a large number of viable sperm cells to fertilize the oocytes. Additionally, the sorted sperm exhibit a reduced survival and increased susceptibility to cryopreservation. Further, the use of sex-sorted sperm by the dairy industry is often limited by the geographical distance between potential sires and the sex-sorting facility.

Appreciable adoption of this technology will depend on whether producers can economically justify the cost to process the sexed-semen. Recent advancements have overcome some of the problems involved with flow cytometry/ cell sorting. New procedures have decreased sorting time by four times, however these rates only allow 6 to 10 straws of semen to be sorted per hour. Deep uterine insemination has lead to a decrease in the number of sperm cells required (2 million versus 15 to 30 million) when inseminating bovine females with sorted semen, allowing for more straws to be processed per hour. Agencies involved in AI are selling sexed bovine-semen to livestock producers on a commercial basis. The cost of sexed-semen ranges from 32-50 USD for proven dairy bulls, while the price is highly variable for beef bulls (http://edis.ifas.ufl.edu/AN163, accessed on Oct. 5, 2009). Problems which remain to be overcome include the cost, genetic quality and availability of oocytes; culture regimes for consistent success in IVM / IVF; and attaining acceptable fertility rates with frozen semen which is thawed and then sorted, and with fresh semen which is sorted and then frozen. In buffaloes, the technology is in early stages of validation but may become a method of choice in future in the developing countries.

Table 1. A summary of applications of sexing in livestock.

1. Choice to have calf of desired sex to promote dairy buffaloes or males for agriculture and meat production
2. Reducing the management costs
3. Faster dissemination of desired females
4. Detection of genetic abnormalities in embryos, and selection of healthy embryos for commercial livestock production system
5. Improving the likelihood of a successful pregnancy and/or viable offspring.
6. Molecular identification and analysis of genes of reproductive importance including identification of candidate genes of embryo developmental competence
7. Accessing the transgenic integration in livestock embryos and selecting the embryos having exogenous gene.
8. Analysis of selected genetic markers at the pre-elongation embryo stage

References

Appa Rao KB, Mohan D and Totey SM. **1994**. Polymerase chain reaction and its applications: special emphasis on its role in embryo sexing. *Biotechnol Adv.* **12**, 341-55.

Blondin P, Beaulieu M, Fournier V, Morin N, Crawford L, Madan P and King WA. **2009**. Analysis of bovine sexed sperm for IVF from sorting to the embryo. *Theriogenology*. **71**, 30-38.

Ceraedts JP and Wert GM. **2009**. Preimplantation genetic diagnosis. *Clin. Genet.* **76**, 315-325.

Cran DG. **2007**. XY sperm separation and use in artificial insemination and other ARTs. *Soc. Reprod. Fertil. Suppl.* **65**, 475-491.

Curry E, Pratt SL, Lapin DR and Gibbons JR. **2009**. Efficacy of a comercially available post-thaw bovine semen sexing kit in both single-ovulating and hyperstimulated cows. *Anim. Reprod. Sci.* **116**, 376-380.

De Vries A, Overton M, Fetrow J, Leslie K, Eicker S and Rogers G. **2008.** Exploring the impact of sexed semen on the structure of dairy industry. *J. Dairy Sci.* **91**, 847-856.

Fulwyler MJ. **1977**. Hydrodynamic orientation of cells. *J. Histochem. Cytochem.* **25**, 781-783.

Fu Q, Zhang M, Qin WS, Lu YQ, Zheng HY, Meng B, Lu SS and Lu KH. **2007**. Cloning swamp buffalo SRY gene for embryo sexing with multiplex-nested PCR. *Theriogenology*. **68**, 1211-1218.

Garner DL and Seidel GE Jr. **2008**. History of commercializing sexed semen for cattle. *Theriogenology*. **69**, 886-895.

Handyside AH, Kontogianni EH, Hardy K and Winston RM. **1990**. Pregnancies from biopsied human preimplantation embryos sexed by Y-specific DNA amplification. *Nature*. **344**, 768-770.

Hirayama H, Kageyama S, Takahashi Y, Moriyasu S, Sawai K, Onoe S, Watanabe K, Kojiya S, Notomi T and Minamihashi A. **2006**. Rapid sexing of water buffalo (*Bubalus bubalis*) embryos using loop-mediated isothermal amplification. *Theriogenolgy*. **66**, 1249-1256.

Liang XW, Lu YQ, Chen MT, Zhang XF, Lu SS, Zhang M, Pang CY, Huang FX, and Lu KH. **2008**. *In vitro* embryo production in buffalo (*Bubalus bubalis*) using sexed sperm and oocytes from ovum pick up. *Theriogenology*, (in press).

Lonergan P. **2007**. State-of-the-art embryo technologies in cattle. *Soc. Reprod. Fertil. Suppl.* **64**, 315–325.

Lu YQ, Zhang M, Meng B, Lu SS, Wei YM and Lu KH. **2006.** Identification of X- and Y-chromosome bearing buffalo (*Bubalus bubalis*) sperm. *Anim Reprod. Sci.* **95**, 158-164.

Lu YQ, Liang XW, Zhang M, Wang WL, Kitiyanant Y, Lu SS, Meng B and Lu KH. **2007**. Birth of twins after *in vitro* fertilization with flow-cytometric sorted buffalo (*Bubalus bubalis*) sperm. *Anim. Reprod. Sci.* **100**, 192-196.

Mara L, Pilichi S, Sanna A, Accardo C, Chessa B, Chessa F, Dattena M, Bomboi G and Cappai P. **2004**. Sexing of in vitro produced ovine embryos by duplex PCR. *Mol. Reprod. Dev.* **69**, 35-42.

Moore K and Thatcher WW. **2006**. Major advances associated with reproduction in dairy cattle. *J. Dairy Sci.* **89**, 1254-1266.

Park JH, Lee JH, Choi KM, Joung SY, Kim JY, Chung GM, Jin DI, Im KS. **2001**. Rapid sexing of preimplantation bovine embryo using consecutive and multiplex polymerase chain reaction (PCR) with biopsied single blastomere. *Theriogenology*. **55**, 1843-53.

Pinkel D, Gledhill BL, Lake S, Stephenson D and Van Dilla MA. **1982a**. Sex preselection in mammals? Separation of sperm bearing Y and "O" chromosomes in the vole *Microtus oregoni*. *Science*. **218**, 904–905.

Pinkel D, Lake S, Gledhill BL, Van Dilla MA, Stephenson D, and Watchmaker G. **1982b**. High resolution DNA content measurements of mammalian sperm. *Cytometry*. **3**, 1-9.

Prashant, Gour DS, Dubey PP, Jain A, Gupta SC, Joshi BK and Kumar D. **2008**. Sex determination in 6 bovid species by duplex PCR. *J. Appl. Genet.* 49, 379-381.

Van den Engh G and Stokdijk W. **1989**. Parallel processing data acquisition system for multilaser flow cytometry and cell sorting. *Cytometry*. **10**, 282–293.

Zhang ZP, Zhang Y, Liu JP, Zhang JT, An ZX, Quan FS, Zhang L, Cai X and Pu SW. **2009**. Codeposition of dNTPs detection for rapid LAMP-based sexing of bovine embryos. *Reprod. Domest. Anim.* **44**, 116-121.

—□□□—

7

Somatic Cell Nuclear Transfer

"The dream of a farmer having an elite animal is to repeat it. Animal cloning seems to ensure this repetition, and a farmer can expect a herd composed of only highly productive genotypes. The technology can expand the reproductive potential of the best animals, allowing breeders to keep up with the demand for semen, embryos and offspring. In addition, the cloning opens a new era in the field of reproductive biology and medicine with respect to the therapeutic cloning"

Introduction

The word clone comes from the Greek, *klon*, meaning a twig or a small branch, and the cloning technique as a scientific procedure has different meanings and protocols in biological sciences. The process of cloning involves asexual reproduction, and is a common method of multiplication in plants and lower invertebrates wherein a fragment of the body can produce an adult animal. However, the higher animals like mammals entirely lack this ability, and reproduce primarily by sexual multiplication.

The first nuclear transfer experiments were performed in 1952 by Briggs and King and were followed by Gurdon to address the fundamental question of whether the genome of different somatic cells are equivalent to those of the cells in early embryos (Briggs and King, 1952; Gurdon, 1962). The principal is that each of the body cells contains full genetic code for producing whole organism, and the nuclear transfer reprograms the adult cells and provides the ways of redirecting the cells to embryo and, subsequently, into an adult animal. Hence, cloning by somatic cell nuclear transfer (SCNT) circumvents the processes that normally function during gametogenesis to enable the gametes to support development of new progeny after fertilization. In more than a decade since the birth of Dolly, the animal cloning efficiency is still frustratingly low, though cloned offspring have been produced in about 18 mammalian species including a number of important farm animal species (Oback 2008; Loi *et al.*, 2007). At the moment, there are around 160 NT cloning laboratories, across 37 countries, of which 75% are working with livestock (cattle, pig, sheep, goat and buffaloes) (Oback and Wells, 2007).

Progress in somatic cell nuclear transfer

The factors involved in the success of SCNT are numerous and very complex. The success of nuclear transfer depends on the source of donor nucleus, developmental stage of recipient cytoplasts, and cell cycle synchrony of donor nucleus. Different methods of harvesting cells have effect on the efficiency of NT. The somatic nucleus will be reprogrammed after NT and will restore a totipotent state in order to undergo development. Although many protocols have been modified and utilized in NT process, some events remain to continue as elusive. The microarray analysis of global gene expression in cloned embryo have unanimously revealed the inefficient reprogramming as major cause of low efficiency of nuclear transfer. However, the number of the mammalian species cloned is increasing, and novel modifications in the technical protocols and reprogramming of the nuclear genome (Wadhwa *et al.*, 2009; Zhao *et al.*, 2009) have come up. Among this is production of cloned calf in Indian Murrah buffalo using handmade cloning, camel (Wani *et al.*, 2009), and birth of mice pups from induced pluripotent stem (iPS) cells (Zhao *et al.*, 2009). A report on SCNT with nuclei derived from mammoth tissues has indicated that it might be possible to promote the cell cycle of nuclei from ancient animal tissues by suitable pre-treatment in SCNT (Kato *et al.*, 2009).

Methodology of nuclear transfer cloning

Preparation of micro tools

The removal of nuclear material requires specialized micromanipulators. Three types of tools, *viz.*, holding pipette for holding the oocyte, cutting needle **(Fig.1)** for cutting the oocyte zona pellucida for extrusion of polar body and nuclear material **(Fig. 2)**, and transfer pipette **(Fig.3)** for transferring the somatic cell into cytoplast are essentially needed. These micro tools are made up of special quality glass capillaries, and can be made in the laboratories or purchased from the commercial manufacturers. Preparation of the tools and their operation requires a technical skill. Alternatively, handmade cloning can be used which avoids the use of the micro tools.

Establishing somatic cell cultures

The somatic cells are used as donor nuclei. For this, the cells have to be cultured *in vitro* and prepared for the subsequent technical steps. Theoretically, any somatic cell can be used as donor nucleus. However, certain cells like embryonic stem cells, granulosa cells **(Fig. 4)**, cumulus cells and fetal somatic cells are reported to be better donors. This is attributed to the ability of the ES cells to be reprogrammed more efficiently following transfer into the cytoplasts.

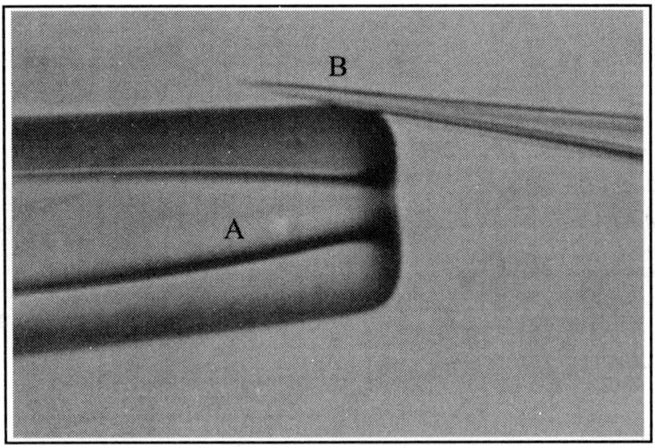

Fig.1. Microtools, A- holding pipette, B- cutting needle

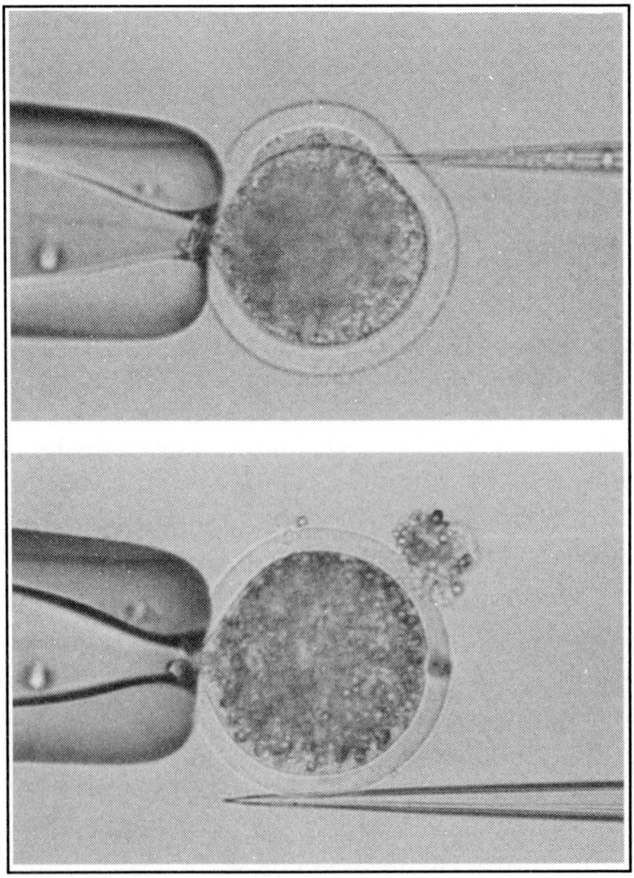

Fig. 2. Nuclear material removal

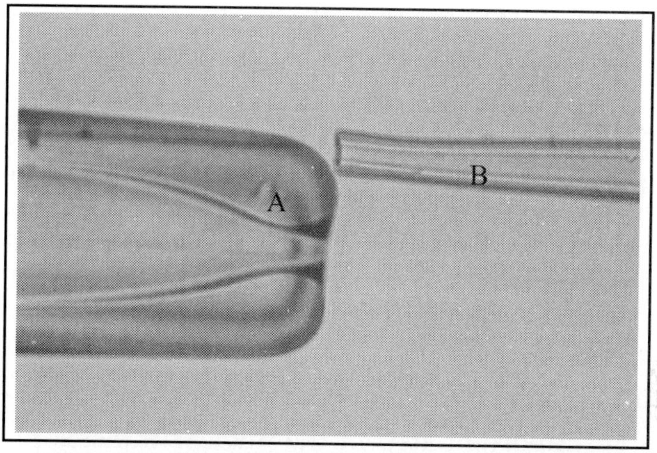

Fig. 3. SCNT holding (A) and transfer pipettes (B)

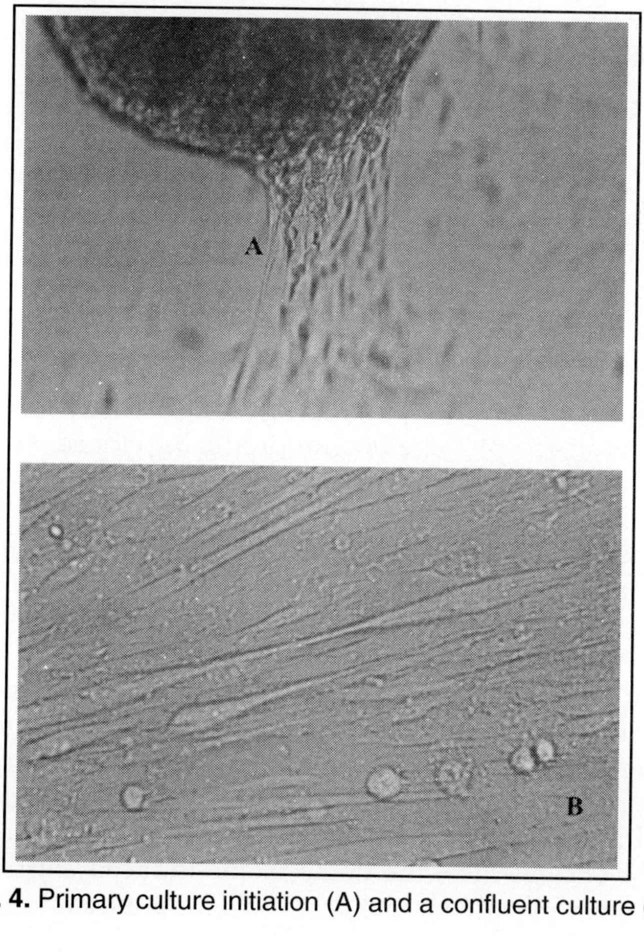

Fig. 4. Primary culture initiation (A) and a confluent culture (B)

Sources, *in vitro* maturation and enucleation of the oocytes

The slaughtered animals are the most common and convenient source of ovaries for obtaining the oocytes. The care is taken to assure the quality of the oocytes. Often overlooked in SCNT, yet the quality of oocyte plays a critical role in the activation and subsequent development of the embryos. In buffaloes, the oocytes or cumulus oocyte complexes (COC) with intact multilayer of cumulus cells and homogeneous granular ooplasm are considered to be good. The oocytes obtained from the live animals can also be used.

Enucleation

This is an important step in the nuclear transfer process. The objective of enucleation is maintaining the ploidy of the organisms to be cloned. Cytoplasts are prepared by removing the nuclear materials of *in vitro* matured oocytes. There are different modifications of the original method of enucleation of the oocytes.

It has been observed that nucleus in the buffalo oocytes is compactly embedded into the lipid-enriched ooplasm and visualization of the pronuclei requires high speed centrifugations (Verma *et al.*, 2006). Nevertheless, the removal of nuclear chromatin from IVM oocytes is hindered by the oocyte lipid contents. Studies have shown that certain chemicals like demecolcine would greatly facilitate the removal of nuclear genetic materials from the IVM oocytes, thereby facilitating the production of cytoplasts.

Derivation of viable cytoplasts through chemically-assisted enucleation is an accepted method for producing cloned embryos, and has been found applicable in farm animals like sheep, cattle (Vajta *et al.*, 2005; Tani *et al.*, 2006) and porcine (Li *et al.*, 2006). It has also been emphasized that induced enucleation performed during meiosis-II could be achieved by a brief exposure of pre-activated oocytes to demecolcine (Baguisi and Overstrom, 2000; Fischer *et al.*, 2002; Ibanez *et al.*, 2003; Gasparrini *et al.*, 2003; Saraiva *et al.*, 2009). The recent approaches of chemically-induced enucleation are based on the phenomenon that treatment of M-II phase with some cytoskeleton-relaxing agents induces an extrusion cone over the surface of the oocyte cytoplasm, which facilitates the removal of genetic material by mechanical aspiration or manual bisection of zona-free oocytes. In buffalo IVM oocytes also, the demecolcine-induced extrusion cones **(Fig. 5)** are more prominent and accessible for removal in most of the zona-included and zona-free oocytes. In some zona-free oocytes, extrusion cones were found to be loosely associated with the oocyte cytoplasm **(Fig. 6)** and a simple gyratory shaking was effective for their detachment or dislocation, leaving behind almost 100% of the cytoplast.

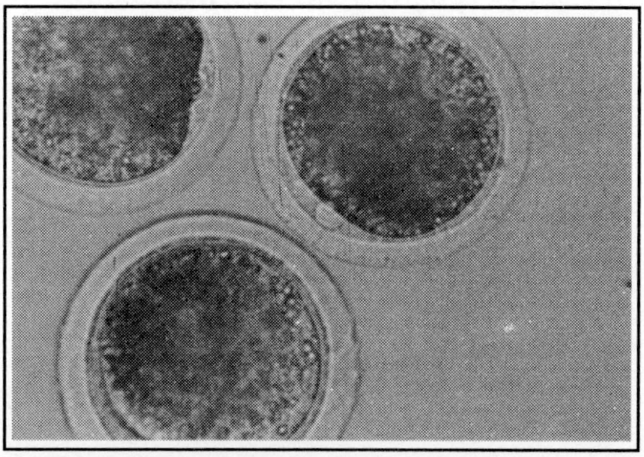

Fig. 5. Extrusion cones in zona-included IVM oocytes

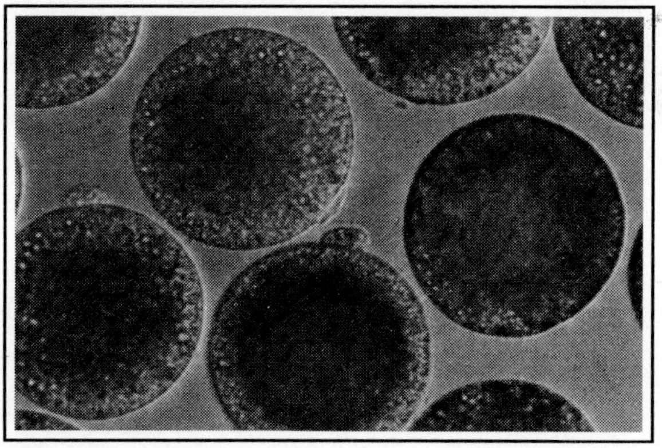

Fig. 6. Extrusion cones in zona-free IVM oocytes

Preparation of cytoplasts

A cytoplast is an oocyte whose nuclear genome has been removed deliberately. The nuclear material of the host oocyte must be removed and replaced with the nucleus of the somatic cell with as little perturbation as possible.

In buffaloes, the immature COCs matured *in vitro* for 20-22h, exhibiting a uniformly expanded and surface anchoring cumulus cells may be used for preparing the cytoplasts (Singh, 2006). The IVM oocytes showing the first polar body are denuded physically by striped-off or by treating them with 0.1% (w/v) trypsin solution, followed by a gentle repeated micropipetting. The enucleation is performed by micromanipulators (e.g. Narshige, Japan; Eppendorf Micromanipulators, North America) attached to an inverted microscope.

In handmade cloning (HMC), the cytoplasts **(Fig. 7)** are prepared by manual bisection of IVM oocytes after enzymatic hydrolysis of the zona pellucida. The handmade enucleation appears to be easier that the conventional micromanipulator-guided enucleation.

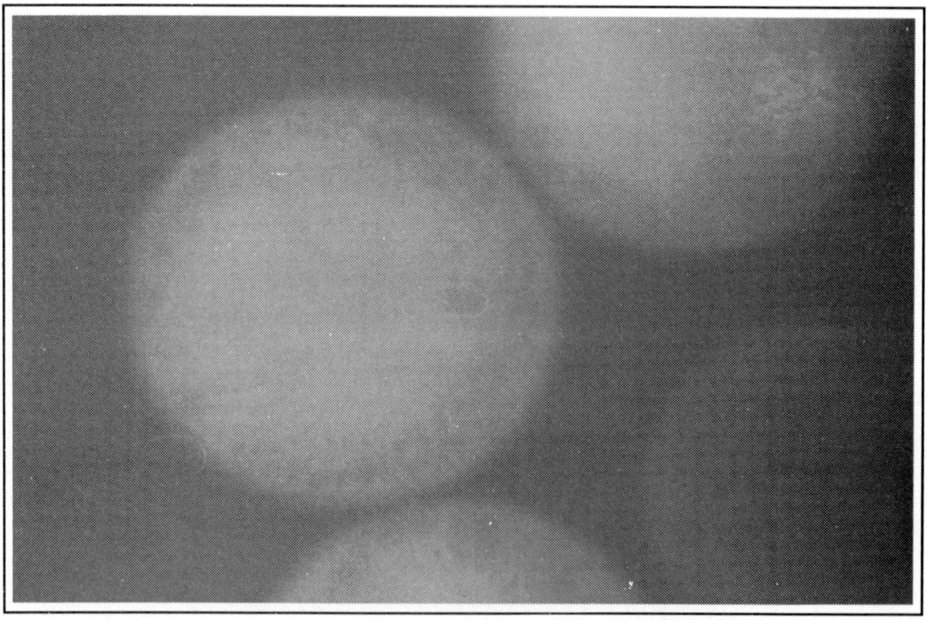

Fig. 7. Cytoplasts

Demecolcine-induced enucleation appears to be promising for preparation of the cytoplast (Baguisi and Overstrom 2000; Gasparrini *et al.*, 2003; Singh 2006; Tani *et al.*, 2006; Shah *et al.*, 2008). Tani *et al.*, (2006) have inferred that demecolcine-treatment for at least 30 min. induces a membrane protrusion in M-II stage bovine oocytes, thus facilitating the enucleation though the treatment had no positive effect on developmental competence of the oocytes. Further, a calf was born from NT embryos produced from demecolcine-treated oocytes, indicating that chemically-induced removal of chromosomes is effective in preparing the cytoplasts and has no adverse effects on cloning (Tani *et al.*, 2006). Similarly, in sheep, a combined treatment with demicolcine and cycloheximide or 6-dimethylaminopurine led to single pronuclear formation rather than PB1 extrusion (Hou *et al.*, 2006).

Certain commercial companies are offering the systems with software for identification of cytoplasts where no cytotoxic UV light is needed and no guess-work is required. A minimum of cytoplasm is removed and enucleation can be confirmed by observing the extracted spindle. This information is available online.

Selection of donor nuclei

Selection of donor nucleus remains an important criterion determining the efficiency of animal cloning. Theoretically, any somatic cell can be cultured for use as donor nuclei. The cells have to be cultured and prepared specifically before fusion with cytoplast. The specialized treatments include serum starvation, treatment with certain agents like trichostatin A, cytochalasin B, or roscovitine etc. However, there is no concluding report about suitability of these methods in all the species. Theoretically, any cell can be cultured and used as donor nuclei, but practically a few types of cells are found to be more suitable as donor nuclei. Among the pioneering studies on production of cloned embryos in buffaloes, fetal fibroblasts were used for production of cloned embryos. It was then also found that granulosa cells can also be reprogrammed for producing the cloned buffalo embryos. Though cumulus cells and granulosa cells are more commonly used donor nuclei in livestock species and laboratory animals, the clones reported in the water buffaloes are from fetal fibroblasts.

Electrofusion of the donor nucleus to the cytoplast

The electrofusion is a critical technical step in NT cloning. The effectiveness of electrofusion of somatic cell into the cytoplast limits the cloning success. In the process, the specialized equipment (e.g. Electrocell Manipulator or Electroporator) is used for generation of an AC wave for benign dielectrophoretic alignment of the cells.

The couplets are placed between the electrodes of cell fusion chamber while visualizing the process under a stereo zoom microscope. The care has to be taken so that the cell-cytoplasts are in a plane perpendicular to the AC waves induced for electroporation of the cells.

With only microsecond switch-over time from AC to DC, efficient fusion can take place. After fusion, AC re-application keeps the cells together for the rounding off process. The DC square pulse section can be used as a stand alone powerful electoporation system with a wide range of voltages and pulse lengths capable of operating into low impedance loads.

Electrofusion has to be optimized to suit for different type and sizes of the cells.

BTX (Harward Apparatus) is at the forefront of electroporation technique. Other companies like Bio-Rad (USA) also supply the machine that can be used for electrofusion of the somatic cell nuclei to the cytoplasts. Virus-mediated cell fusion is also reported by certain groups, but the use of viruses is discouraged because of risks involved in handling the viruses.

The critical steps to be taken into consideration include magnitude of the impulse needed for alignment of the orientation of somatic cell adhered to the cytoplast (in HMC) or the alignment of the somatic cells transferred into the perivitelline membrane of the cytoplast as in traditional cloning. The researchers have used different combinations of the AC and DC to perform electrofusion.

In vitro culture of the cloned embryos

It is necessary that the newly constructed couplets be examined for viability. The quiescent couplets are removed and only actively dividing ones are cultured further. Generally, the embryos reaching hatching stage of blastocyst **(Fig. 8)** are found suitable for transfer. The process needs a suitable combination of culture media and culture conditions. Both the media prepared in the laboratory or ready-to-use culture media are used by various researchers. Based on the reports on production of cloned embryos mCR2aa and mSOFaa are found to support *in vitro* growth of the cloned embryos. However, to avoid batch to batch variation, ready-to-use culture media should preferably be used. The embryo culture medium employed in the production of cloned embryos by Shi *et al.* (2007) comprised of TCM-199 supplemented with 3% estrus cow serum (ECS), penicillin G (60 μg mL^{-1}), and streptomycin (100 μg mL^{-1}). The medium was filter-sterilized and stored at 4°C for 4 weeks. Another research group from National Dairy Research Institute (India) (Shah *et al.*, 2008) reported that culture media used in pre-implantation development of handmade cloned embryos included modified Charles Rosencrains 2 medium with amino acids (mCR2aa), modified synthetic oviductal fluid with amino acids with citrate and myoinositol (mSOFaaci) (Holm *et al.*, 1999). All the culture media were supplemented with fatty acid-free BSA. Third culture medium was ready-to-use medium named RVCL (Research Vitro Cleave medium, Cook ®, Australia). The media, namely mCR2aa, mSOFaa and RCVL yielded nil (0), 4.6 $\pm$ 1.8 (7) and 40.3 $\pm$ 4.7 (50) (mean $\pm$ SEM (n)) embryos, respectively. The pre-implantation developmental competence was better on the flat surface compared to well-of-the wells (WOW) or MD systems (Shah *et al.*, 2008). A modified sequential media method, named D medium (M16/KSOM and CZB-EG/KSOM), was successfully established that significantly improves SCNT embryo development (Dai *et al.*, 2009). The results demonstrated that D medium did not have any apparent adverse effects on *in vitro* fertilized embryos and dramatically improves the blastocyst development of SCNT embryos compared with other commonly used culture media including KSOM, M16, CZB, and αMEM. Specifically, the rate of blastocyst formation was 62.3% for D1 (M16/KSOM) *vs.* 10-30% for other media. An analysis of media components indicated that removing EDTA and glutamine from the media can be beneficial for early SCNT embryo development (Dai *et al.*, 2009).

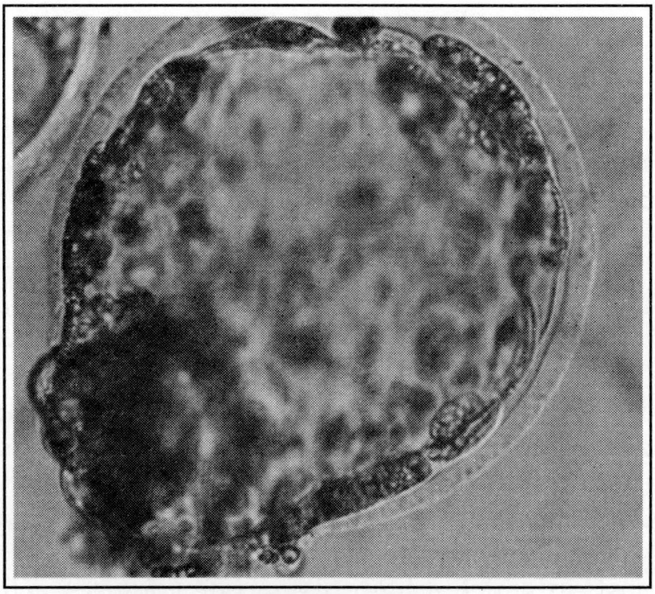

Fig. 8. Cloned embryo

Estrus synchronization and transfer of the cloned embryos

Synchronization of estrus refers to the process of scheduling a group of recipient animals to come into estrus at approximately the same time as donor animals which have been superovulated with exogenous drugs administration. The care has to be taken that the females being used as surrogate mothers are reproductively efficient, irrespective of their milk yield. Generally low-yielding but reproductively efficient females are selected as surrogate mothers, whereas high-yielding females are used as oocyte-donors. A number of commercially available drugs such as prostaglandin PGF-2α (InterVet France) and combinations of prostaglandin and gonadotropins are used for estrus-synchronization. The efficacy of the drugs and the success of application protocols may vary depending on the time of application, nutrition and physiological status of the animals.

The embryos are transferred to the surrogate mothers but the birth of a clone depends on several factors including accurate temporal expression of the key genes associated with pluripotency in early-implantation and post-implantation periods, surrogate mother health, management, and neonatal care.

Cloning in water buffaloes

The studies on producing cloned offspring in buffalo commenced about a decade ago with the reports that *in vitro* cultured fibroblasts and granulosa cells could be used as donor nuclei (Parnpai *et al.*, 2000). Meanwhile, the efforts to increase the efficiency of bubaline embryo cloning remained one of the prime research focuses, though no cloned calves could be produced. Shi *et al.* (2007) for the first time announced births of three female cloned calves, showing that fetal fibroblasts or granulosa cells could be used as donor nuclei for producing cloned progeny. The success was attributed to three major technical aspects. First, a non-invasive enucleation method was employed with the 'Spindle View System' for ensuring the formation of quality cytoplasts. Second, aphidicolin plus serum starvation was employed to synchronize the donor cell cycle at the G0/G1 phase without resulting in damage to the nuclear genome in comparison with serum starvation alone, which may favor the development of reconstituted embryos. Third, the electrofusion method was applied to fuse the donor cell and cytoplast, in which a high fusion rate could be achieved in a relatively low intensity of electric field (Shi *et al.*, 2007).

As an alternative to conventional SCNT, the handmade cloning (HMC) has been reported to improve bubaline embryo cloning for larger scale-production of the cloned embryos (Singh *et al.*, 2008; Shah *et al.*, 2008), and establishing pregnancies in buffaloes (Shah *et al.*, 2009). Births of cloned calves from the transfer of zona-free or HMC embryos in Indian Murrah buffalo are reported recently in India (http://www.icar.org.in/news/World-First-Cloned-Buffalo-Calf-Born.htm, accessed Feb. 2, 2009), which died a few days after. In June 2009, another cloned calf "Garima" **(Fig. 9)** has been reported from the same laboratory (b2bportal.ohlog.com/**garima**-india.oh66736.html, accessed Nov. 04, 2009). Garima's birth is a testimony to India's low-cost scientific prowess. The success has now raised unprecedented hopes for conservation and genetic improvements of dairy buffaloes, and to produce transgenic cloned buffaloes for production of therapeutic products. As our understanding of epigenetics and assisted reproduction is incomplete, much remains to be investigated about the fundamental processes and mechanisms that can promote the animal cloning for commercial applications.

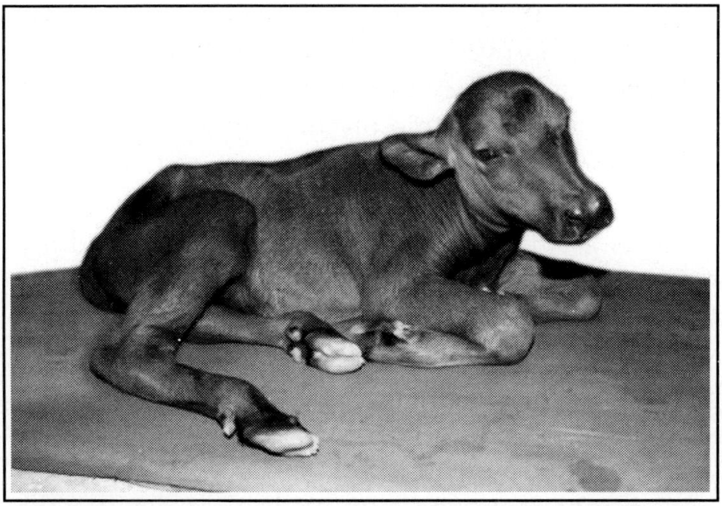

Fig. 9. Garima, the second handmade cloned buffalo calf

Opportunities and challenges

Animal cloning is becoming increasingly useful for its applications in biological inquiry and for its potential use in pharmaceutical, medical and agricultural fields. Despite enormous scientific inputs and experience of cloning of as much as 20 mammalian species, cloning by nuclear transfer is often associated with poor results primarily due to aberrant nuclear reprogramming of somatic donor cells and altered gene expression patterns. Widely reported problems with animals derived from nuclear transfer embryos concern the abnormalities of general and specific tissue growth both *in utero* and after birth. The experience included extended gestation, increased still-birth rate and perinatal deaths.

Abnormalities in the expression of imprinted genes are major causes of the low efficiency of SCNT procedures. Moreover, in view of the complexity of the steps required in reconstructing oocytes by nuclear transfer, new strategies should be evolved to minimize the developmental damages and improve the reprogramming ability **(Table-2)**. High embryonic and fetal attrition rates of cloned animals are commonly observed in ruminants that seem to be particularly sensitive to epigenetic aberrations. In addition, cardiopulmonary and placental lesions are also observed. The abnormalities appear to be related to the cloning procedures *per se*, and to some extent the neonate and surrogate mother care. Better selection and pre-treatment of donor nuclei and use of improved culture media and techniques during the pre-implantation stage, may hopefully improve the technology for practical applications in buffalo production system.

Conclusion

Cloning is a valuable tool in any strategic breeding program because it can more quickly and precisely extend elite genotypes for additional embryo or semen production, increased breeding opportunities, protection from untimely loss of superior individuals. By cloning the most elite animals, with the most sought-after genetics or a higher disease-resistance, the livestock breeders can advance the efficiency, quality and consistency of their livestock products. Another important insight from the cloning is that despite decrease in developmental potential the nucleus of most, if not all, adult cells retains nuclear plasticity and can be reset to an embryonic state. Due to the complexity of the numerous steps required in reconstructing oocytes by nuclear transfer, novel protocols are required to minimize the damages inflicted during these manipulations. Moreover, it is essential that protocols be adapted according to each species concerned as the oogenesis and early embryogenesis differ widely among mammalian species. Efficiency of animal cloning can be greatly improved using optimized operational procedures, including treating the donor cells, appropriate activation of fused embryos, use of suitable culture medium for pre-implantation development of the embryos, improved management of surrogate mothers, and neonate healthcare.

Table 1. A comparison of conventional nuclear transfer and handmade cloning techniques in respect of technical requirements

Traditional nuclear transfer cloning (SCNT)	Handmade cloning (HMC)
Needs microotools, *viz.*, holding microtube and cutting needle	No need of microtools. It needs a specialized ultrasharp cutting blade
Requires micro tool preparing equipment and micromanipulator to prepare cytoplast and donor cell transfer	Not required
Technically complex and expensive	Simpler and cheaper
The overall operation efficiency is slow	It is comparatively faster

Table 2. A summary of suggestions to improve SCNT for applications in livestock

1. Use of oocytes from the live donors may significantly improve the SCNT efficiency
2. Using HMC, instead of conventional SCNT will prove to be cheaper, efficient and convenient for reconstituting the couplets
3. Preparing cytoplasts using chemically-assisted enucleation
4. Studies should focus on identification of molecular markers as criteria for selection of the donor nuclei
5. Somatic cell banks of the candidate donor nuclei should be established to increase the efficiency of the technique
6. Emphasis should be on identifying a single or minimum number of culture media for improving the pre-implantation developmental competence of the cloned embryos
7. Improved care of surrogate mother carrying the nuclear transfer cloned fetus
8. Proper health care of neonate cloned calves

Table 3. Summary of the technical recommendations based on the review of literature to improve SCNT efficiency

Treatment	Mode of action (reference)
Donor nuclei	Selection of the certified somatic cells as candidate donor nuclei; treatment of donor nuclei with pronase to achieve higher elcetrofusion efficiency (Zhao et al., 2009); treating the donor nuclei with trichostatin A (Cervera *et al.*, 2009) and roscovitine; exploring the use of ES cells or other pluripotent cells as donor nuclei.
	Addition of chromatin remodeling factors to the oocyte for facilitating reprogramming of the cell nucleus (Betthauser JM *et al.*, 2006)
Enucleation	Using handmade cloning instead of SCNT to achieve higher efficiency of cytoplast generation, minimizing the exposure to hazardous chemicals and irradiation, increased rate of somatic cell fusion to the cytoplast within minimum time period
Electrofusion and activation	Using an appropriate combination of chemical agents for activation of the electrofuced embryos (Zhang *et al.*, 2008); using appropriate culture medium (Shah *et al.*, 2008); exploring the alternative options for improving the nuclear reprogramming (Bui *et al.*, 2008)
Improved embryo culture	Novel *in vitro* culture system for improving pre-implantation embryo developments, e. g. microfluidic *in vitro* cultivation system (Kim *et al.*, 2009); Well-of-the-Well system (Vajta *et al.*, 2005, 2008) etc.

References

Baguisi A and Overstrom EW. **2000**. Induced enucleation in nuclear transfer procedures to produce cloned animals. *Theriogenology*. **54**, 209. (Abstract)

Betthauser JM, Pfister-Genskow M, Xu H, Golueke PJ, Lacson JC, Koppang RW, Myers C, Liu B, Hoeschele I, Eilertsen KJ, Leno GH. **2006**. Nucleoplasmin facilitates reprogramming and in vivo development of bovine nuclear transfer embryos. *Mol. Reprod. Dev.* **73**, 977-986.

Briggs R and King TJ. **1952**. Transplantation of living nuclei from blastula cells into enucleated frogs' eggs. *Proc. Natl. Acad. Sci. U S A*. **38**, 455–463.

Bui H, Wakayama S, Kishigami S, Kim J, van Thuan N and Wakayama T. **2008**. The cytoplasm of mouse germinal vesicle stage oocytes can enhance somatic cell nuclear reprogramming. *Development*. **135**, 2935-2945.

Cervera RP, Marti-Gutierrez N, Escorihuela E, Moreno E and Stojkovic M. **2009**. Trichostatin A affects histone acetylation and gene expression in porcine somatic cell nuclear transfer embryos. *Theriogenology*. **72**, 1097-1110.

Dai X, Hao J and Zhou Q. **2009**. A modified culture method significantly improves the development of mouse somatic cell nuclear transfer embryos. *Reproduction*. **138**, 301-308.

Fischer D, Ibanez E, Cibelli JB, Albertini DF, Overstrom EW. **2002**. Activated bovine cytoplasts produced by induced-enucleation support development of bovine nuclear transfer embryos. *Biol. Reprod.* **66 (Suppl.1)**, 238.

Gasparrini B, Gao S, Ainslie A, Fletcher J, McGarry M, Ritchie WA, Springbett AJ, Overstrom EW, Wilmut I and De Sousa PA. **2003**. Cloned mice derived from embryonic stem cell karyoplasts and activated cytoplasts prepared by induced enucleation. *Biol. Reprod.* **68**, 1259-1266.

Gurdon J. **1962**. The developmental capacity of nuclei taken from intestinal epithelium cells of feeding tadpoles. *J. Embryol. Exp. Morphol.* **10**, 622–640.

Holm P, Booth PJ, Schmidt MH, Greve T and Callesen H. **1999**. High bovine blastocyst development in a static in vitro production system using SOFaa medium supplemented with sodium citrate and myo-inositol with or without serum proteins. *Theriogenology*. **52**, 683-700.

Ibáñez E, Albertini DF, Overström EW. **2003**. Demecolcine-induced oocyte enucleation for somatic cell cloning: coordination between cell-cycle egress, kinetics of cortical cytoskeletal interactions, and second polar body extrusion. *Biol. Reprod.* **68**, 1249-58.

Hou J, Lei T, Liu L, Cui X, An X, Chen Y. **2006**. Demecolcine-induced enucleation of sheep meiotically maturing oocytes. *Reprod. Nutr. Dev.* **46**, 219-26.

Kato H, Anzai M, Mitani T, Morita M, Nishiyama Y, Nakao A, Kondo K, Lazarev PA, Ohtani T, Shibata Y, Iritani A. **2009**. Recovery of cell nuclei from 15,000 years old mammoth tisuuses and its injection into mouse enucleated matured oocytes. *Proc. Jpn. Acad. Ser. B Phys. Biol. Sci.* **85**, 240-7.

Kim MS, Bae CY, Wee G, Han YM, Park JK. **2009**. A microfluidic in vitro cultivation system for mechanical stimulation of bovine embryos. *Electrophoresis*. **30**, 3276-82.

Li J. Du Y, Zhang YH, Kragh PM, Purup S, Bolund L, Yang H, Xue QZ and Vajta G. **2006**. Chemically assisted handmade enucleation of porcine oocytes. *Cloning Stem Cells*. **8**, 241-250

Loi P, Beaujean N, Khochbin S, Fulka J Jr, Ptka G. **2007**. Asymmetric nuclear reprogramming in somatic cell nuclear transfer? *BioEssays*. **29**, 1-10.

Madan ML, Das SK and Palta P. **1996**. Application of reproductive technology to buffaloes. *Anim. Reprod. Sci.* **42**, 299-306.

Oback B. **2008**. Climbing mount efficiency- small steps, not giant leaps towards higher cloning success in farm animals. *Reprod. Domest. Anim.* **43**(Suppl. 2), 407-416.

Oback B and Wells DN. **2007**. Cloning cattle: methods in madness. Review. *Adv. Exp. Med. Biol.* **591**, 30-57.

Parnpai R, Tasripoo K, Kamonpatana M. **2000**. Developmental potential of cloned bovine embryos derived from quiescent and non-quiescent adult ear fibroblasts after different activation treatments. *Theriogenology*. **53**, 239. [Abstract]

Saraiva NZ, Perecin F, Méo SC, Ferreira CR, Tetzner TA, Garcia JM. **2009**. Demecolcine effects on microtubule kinetics and on chemically assisted enucleation of bovine oocytes. *Cloning Stem Cells*. **11**, 141-152.

Shah RA, George A, Singh MK, Kumar D, Chauhan MS, Manik RS, Palta P, Singla S K. **2008**. Hand-made cloned buffalo (*Bubalus bubalis*) embryos: comparisom of different culture media and culture sysytems. *Cloning Stem Cells*. **10**, 435-442.

Shah RA, George A, Singh MK, Kumar D, Anand T, Chauhan MS, Manik RS, Palta P and Singla SK. **2009**. Pregnancies established from handmade cloned blastocysts reconstructed using skin fibroblasts in buffalo (*Bubalus bubalis*). *Theriogenology*. **71**, 1215-1219.

Shi D, Lu F, Wei Y, Cui K, Yang S, Wei J, Liu Q. **2007**. Buffalos (*Bubalus bubalis*) cloned by nuclear transfer of somatic cells. *Biol Reprod*. **77**, 285-291.

Singh AK, Singh B, Singla SK. **2008**. Studies on handmade somatic cells cloning in buffalo (*Bubalus bubalis*). **In**: *International conference on emerging infectious diseases of animals and biotechnological applications*, held at Tamilnadu Veterinary and Animal Sciences University, Chennai (India), 28-29th July 2008, pp. 131. [Abstract].

Singh AK. **2006**. Studies on handmade cloning in buffaloes. M. V. Sc. Thesis submitted to Dairy Science College, National Dairy Research Institute (Deemed University), Karnal. India.

Tani T, Shimada H, Kato Y and Tsunoda Y. **2006**. Demecolcine-assisted enucleation for bovine cloning. *Cloning Stem Cells*. **8**, 61-66.

Vajta G, Maddox-Hyttel P, Skou CT, Tesirlioglu RT, Peura TT, Lai L, Murphy CN, Prather RS, Kragh PM and Callesen H. **2005**. Highly efficient and reliable chemically assisted enucleation method for handmade cloning in cattle. *Reprod. Fertil. Dev.* **17**, 791-797.

Vajta G, Korösi T, Du Y, Nakata K, Ieda S, Kuwayama M and Nagy ZP. **2008**. The Well-of-the-Well system: an efficient approach to improve embryo development. *Reprod. Biomed. Online.* **17**, 73-81.

Verma V, Gautam SK, Singh B, Manik RS, Singla SK, Palta P, Chauhan MS. **2006**. Visualization of pronucleus for transgenic embryo production in buffalo (*Bubalus bubalis*). **In**: *International Congress on Gamete Biology: Emerging Frontiers in Fertility and Contraceptive Development*, organized by NII, New Delhi (February 22-25, 2006). pp 76.

Wadhwa N, Kunj N, Tiwari S, Saraiya M and Majumdar SS. **2009**. Optimization of embryo culture conditions for increasing efficiency of cloning in buffalo (*Bubalus bubalis*) and generation of transgenic embryos via cloning. *Cloning Stem Cells.* **11**, 387-395.

Wani NA, Wernery U, Hassan FA, Wernery R and Skidmore JA. **2009**. Production of the first cloned camel by somatic cell nuclear transfer. *Biol. Reprod.* (in press).

Zhao LW, Yang XY, Guan PF, Fu J, Li H, Zhou YY, Huang SZ, Zeng YT, and Zeng FY. **2009**. Improved efficiency of bovine somatic cell nuclear transfer by optimizing operational procedures. *J. Reprod. Dev.* **55**, 542-546.

Zhao XY, Li W, Lv Z, Liu L, Tong M, Hai T, Hao J, Gulo, CL, Ma QW, Wang L, Zeng F and Zhou Q. **2009**. iPS cells produce viable mice through tetraploid complementation. *Nature.* **461**, 86-90.

Zhang G, Sun QY, Chen DY. **2008**. In vitro development of mouse somatic nuclear transfer embryos: effects of donor cell passages and electrofusion. *Zygote.* **16**, 223-227.

—□□□—

8

Stem Cell Technology

"Embryonic stem cells are of potentially big value in transgenic animal research and studies of lineage commitment and development. Unfortunately, despite many efforts, scientifically validated embryonic stem cell lines in animal species other than mice and primates are not established. A number of technical hurdles have to overcome before the full potential of stem cell research both in livestock applications is realized"

Introduction

Since the beginning of 20th century, scientists have come to understand that all of the different types of blood and tissue cells in the body develop from what have come to be referred to as "stem cells." According to the National Institute of Health (NIH), stem cells have the "remarkable potential to develop into many different cell types in the body". Serving as a sort of repair system for the body, these cells in most multi-cellular organisms can theoretically divide without limit to replenish other cells as long as an animal is alive. The recent upsurge in publication on animal stem cells can be taken as reflection of the interest aroused by these cells due to unique biological properties and numerous applications (Munoz *et al.*, 2009).

Historically, the stem cells research started in 1981 with the milestone publications reporting the establishment of murine embryonic stem cells (Evans and Kaufman, 1981; Martin , 1981). The award of 2007 Nobel Prize in Physiology or Medicine to Drs Capecchi, Evans, and Smithies for their pioneering work on engineering the genetically modified mice (Vogel, 2007) put the stem cells in scientific limelight.

The stem cells are capable of maintaining the ability to multiply meiotically and differentiate into a diverse range of specialized cells types. The stem cells are categorized into two broad areas namely, embryonic stem cells and adult stem cells. In addition, the stem cell exhibiting stem cell-like characteristics can also be derived from the fetus, cord blood and amniotic fluid of the mammalian species.

Embryonic stem (ES) cells are derived from the preimplantation embryos at the time when they reach blastocyst stage, and are characterized by essentially unlimited self-renewal, and exhibit the ability to differentiate into all cell types in an individual. It is at this point that first step of differentiation takes place during mammalian embryonic development. The individual blastomeres now start to organize themselves into three distinct locations, namely, trophectoderm, the primitive endoderm and inner cell mass (ICM) cells. ES cells originate from primitive endoderm, which is a transiently existing group of cells in the embryo. *In vitro* produced blastocysts are commonly used as the starting material for ES cell lines.

Characteristics of ES cells

ES cells are the pluripotent cells with the capacity for long-term propagation and broad differentiation plasticity, and were first established as cell lines from the inner cell mass (ICM) of mouse blastocysts in early 1980s. These cells are characterized by their unique functional feature that upon combination with a host embryo (chimera production) they can contribute to all tissues and organs, including germ cells of the resulting chimeric mouse (Talbot and Blomberg, 2008). As these cells are competent to form all cell types including extraembryonic placental tissues, they are considered totipotent or pluripotent depending on the particular cell line or in environmental context. Furthermore, they can be clonally propagated and maintained in culture indefinitely. These characteristics have made them an invaluable genetic engineering tool for studying functional mammalian genetics, mammalian developmental biology, and for producing animal models of human diseases.

ES cells are isolated from ICM cells of the blastocycts **(Fig.1)**. Different techniques, *viz.* mechanical separation, immunosurgery, and laser-assisted isolation of the ICM cells are employed for isolation of these cells. ES cell lines have similar cell culture properties regardless of the species of origin or the tissue of origin i.e., derivation from morula stage embryos, the ICM of the blastocyst, primordial germ (PG) cells of the embryonic genital ridge, or the early post-implantation epiblast (Brons *et al.*, 2007; Tesar *et al.*, 2007). The murine ES cells typically grow in compact colonial groups, or 'nests' of cells that often have a convex 3-D shape and a distinct, glistening edge that meets with the flatter feeder cells (Robertson, 1987). The ES cells generally grow on top of or in between the feeder cells. Mouse ES (mES) cell colonies grow quickly to hundreds if not thousands of cells per colony, and the colonies will eventually fuse with one another to form monolayers. If left undisturbed the ES cells tend to spontaneously differentiate at the periphery of the colony with the formation of flatter, larger, and irregularly cuboidal visceral endoderm. Morphology of primate ES cell colony is different from mES cells. Human ES cells are generally flatter in appearance and their spontaneous differentiation tends to begin in the center of colonies if they are left undisturbed for a week or more without passage (Thomson and Marshall, 1998; Thomson *et al.*, 1998).

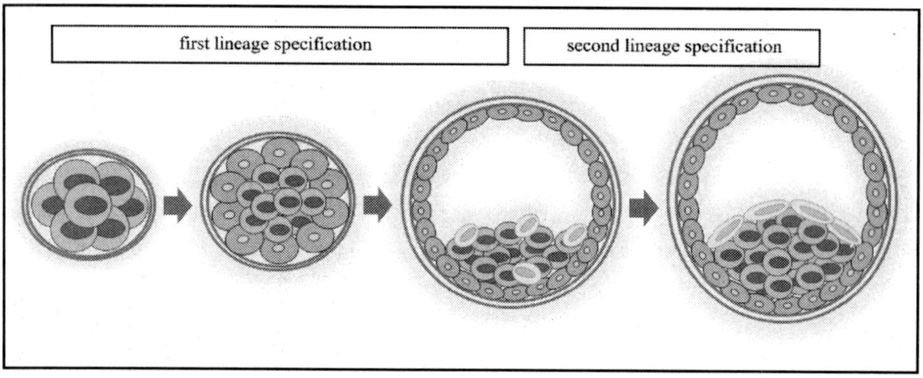

Fig.1. First differentiation in totipotent embryo

Types of stem cells

Adult stem cells and spermatogonial stem cells

Like human beings and other mammalian species, the sources for pluripotent stem cells having differentiation potential to all the three germ layers, are either embryos or germ line **(Fig. 2)**. Postnatal sources of pluripotent stem cells include bone marrow, blood, and skeletal muscles, adipose tissue, skin, fetal somatic explants, umbilical cord, placental tissue and amniotic fluid, etc. **(Fig. 2 and 3)**.

Adult stem cells inhabit specific tissue groups and can both renew themselves and differentiate into the specialized cell type of the tissue in which they reside. For example, hematopoietic stem cells are the cells that can differentiate into any of the different blood cells, while mesenchymal stem cells can do the same for the cells of skeleton. In addition, the adult stem cells may also differentiate into cells of other tissue types, a phenomenon known as transdifferentiation or plasticity.

Spermatogonial stem cells originate from the primordial germ (PG) cells the progenitor cells for both the male and female germ line which, in turn, is derived from the epiblast cells (Lawson and Pederson, 1992). PG cells are the single cells that can under certain culture conditions form colonies of cells, which morphologically resemble undifferentiated ES cells. The spermatogonial stem cells are the only adult stem cells that donate genetic material to the progeny. In bovines, the spermatogonial stem cells can be extracted, cultured *in vitro* and used for IVF or ICSI for producing the embryos. Another broader area of the applications of these cells is their application in developing the transgenic animals.

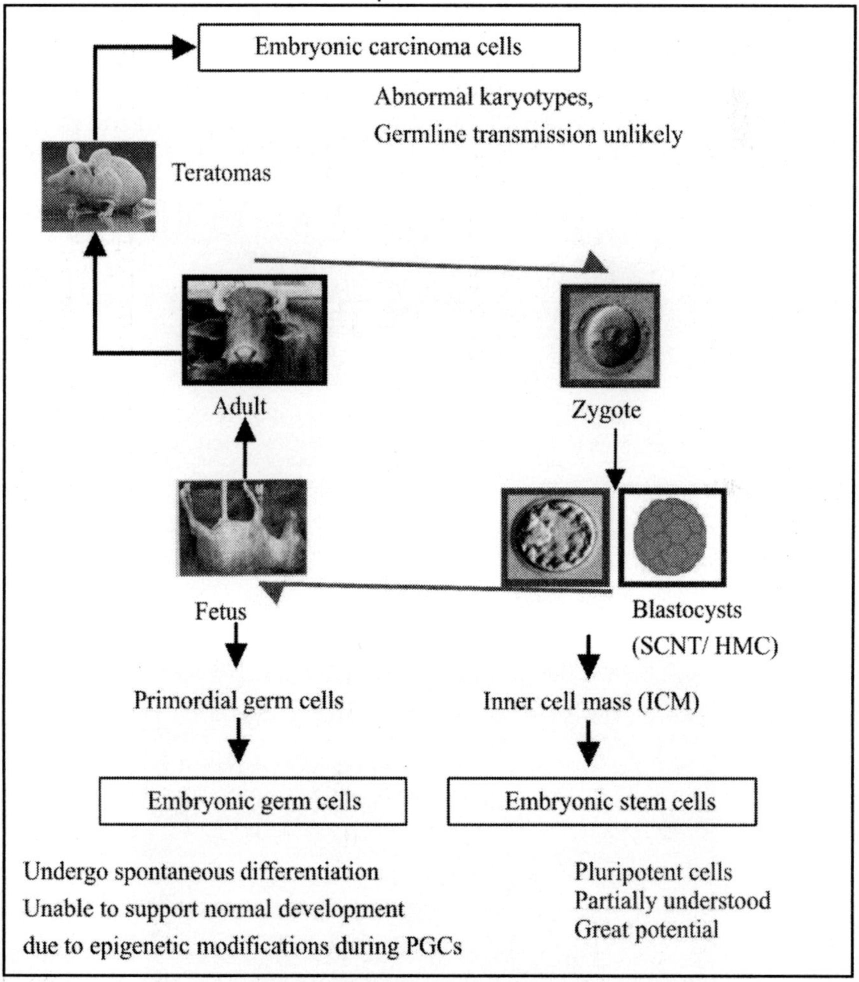

Fig. 2. Sources of stem cells

Recently, novel stem cell type- epi-stem cells (epiSCs) has been the subject of enormous scientific inputs (Brons *et al.*, 2007; Tesar *et al.*, 2007). Under the present status of research, epiSCs derivation in farm animals may prove easier than true ES cell derivation, and could represent a significant step forwards for identifying the factors required for maintaining pluripotency in cultured ICM cells of farm animal embryos (Galli and Lazzari 2008). Nevertheless, the key aspects of animal ES cells, specifically the identification of species-specific ES cell-markers need to be elucidated. It has been pointed out that currently used human or murine ES cell-specific molecular markers are not specific for bovine ES cells, hence, further research is warranted on identification of valid ES cell markers in bovines (Munoz *et al.*, 2008). It was also suggested that until validated pluripotent bovine stem cell markers are identified, it might be advisable to combine the use of epiblast and trophoblast-specific markers to rule out the presence of trophectoderm cells in ES cell cultures (Munoz *et al.*, 2008).

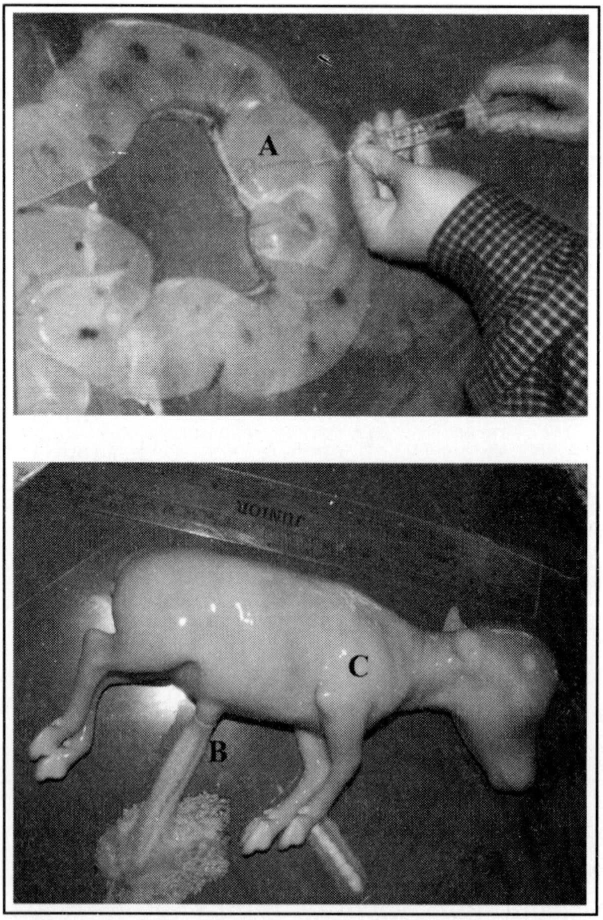

Fig. 3. A- amniotic fluid collection, B- umbilical cord and C- fetal tissue explants.

Introduction and integration of the exogenous genetic material is one of the major challenges in stem cells if they are used for producing transgenic cells or embryos. In recent years, stem cells and nanotechnologies are gradually intercrossing to form a new emerging field, that is, nanotechnology in stem cells or stem cell nanotechnology (Wang *et al.*, 2009). The studies of interaction between nanomaterials or nanostructures and stem cells have made great advances (Wang *et al.*, 2009). Moreover, the innovations in mechanisms of intracellular delivery and integration of foreign genes into the genome through highly efficient, safe and suitable vectors like baculoviral vectors (Chuang *et al.*, 2007; Lufino *et al.*, 2008; Lo *et al.*, 2009), or multifunctional nanoparticles, liposomes or biodegradable polymeric vectors may hopefully develop novel stem cell technologies and advance the existing concept of '*biopharming*' by using buffalo or cattle mammary gland as live bioreactors. The animal genetic engineering along with advances in cell culture technology may be particularly useful when used together and the achievement of the goal seems closer in future.

Mammary gland stem cells

Adult mammary stem cells are multipotent cells which are committed to give rise to cells with a specific function. These cells produce a lineage of daughter cells with a unidirectional terminal differentiation process. Mammary gland adult stem cells are slow-cycling cells with the ability to respond to environmental cues and propagate additional stem cells or differentiate along a specific cell lineage. Once the stem cell commits to differentiation, it enters a brief period of rapid proliferation. Asymmetrical cell division generates one daughter stem cell and one progenitor cell. This process allows for the maintenance of a stem cell pool for future growth while simultaneously generating a differentiated cell population. On the other hand, symmetrical cell division produces two daughter stem cells. This enlarges the stem pool from one cell to many. The murine mammary stem cells (MaSC) have been isolated using methods as employed to identify the hematopoietic stem cells. The isolation methods involve the depletion of hematopoietic and endothelial cells from freshly dissociated cell preparations and differentiation of mammary cell subpopulations based on their cell-surface expression of CD24 (heat-stable antigen), CD29 (β_1-integrin) or CD49f (α_6-integrin) molecules. Interestingly, these integrins are likely to form a functional α_6-β_1 integrin heterodimeric complex that mediates interaction between epithelial cells and the mammary stroma.

Cord blood and amniotic stem cells

In the late 1980s, umbilical cord blood (UCB) was recognized as an important clinical source of HSCs. Blood from the placenta and umbilical cord is a rich source of hematopoietic stem cells, and these cells are typically discarded after the birth. Increasingly, UCB is harvested, frozen, and stored in cord blood banks either as an individual resource (donor-specific source) or as a general resource, directly available when needed. Advantages of cord blood include its availability, ease of harvest, and the reduced risk of graft-versus-host-disease (GVHD). In addition, cord blood HSCs have a greater proliferative capacity than adult HSCs. *Ex vivo* expansion in tissue culture, to which cord blood cells are more amenable than adult cells, is another approach under active investigation. Umbilical cord matrix cells can also be a good source of fetal stem cells. These cells can be isolated and cultured. For this umbilical cord (UC) is collected from new born calves immediately after birth and processed for deriving hematopoietic cells from umbilical cord vein, and mesenchymal cells from Wharton Jelly by using standard cell culture protocols.

Amniotic fluid and cord blood, long considered a waste product, offer considerable therapeutic utility in human beings, but in livestock these stem cells offer a remarkable potential in the genetic engineering of the animals. Umbilical cord blood is extensively rich in hematopoietic stem cells. The amniotic fluid contains the cells which exhibit stem cells like characteristics. The initial studies on culturing and partial characterization of the buffalo amniotic cord blood and amniotic fluid cells **(Fig. 4)**, have raised the hopes of authentic characterization and derivation of somatic cells from these cells.

Induced pluripotent stem cells

At the 2006 meeting of the International Society for Stem Cell Research in Toronto, Ontario, Canada, Shinya Yamanaka announced the startling discovery that adult skin cells could be directly reprogrammed to pluripotency using a combination of only four genes. The project began by making a list of 24 known pluripotency-associated genes, predominantly those expressed in ES cells. Various combinations of these genes were introduced to adult mouse skin fibroblasts via engineered retroviruses in order to screen for gene combinations that could reprogramme the skin cells to pluripotency. Yamanaka was able to define a rather 'simple' cocktail of just four factors namely, Oct4, Sox2, c-Myc and Klf4 that were capable of reprogramming skin cells to a pattern of gene expression that was amazingly similar to mouse ES cells (Takahashi and Yamanaka, 2006). This method was termed 'direct reprogramming' as it relied on primary genetic modulation to force reprogramming instead of NT. The resulting cells were dubbed 'induced pluripotent stem' or iPS cells (Fig.5).

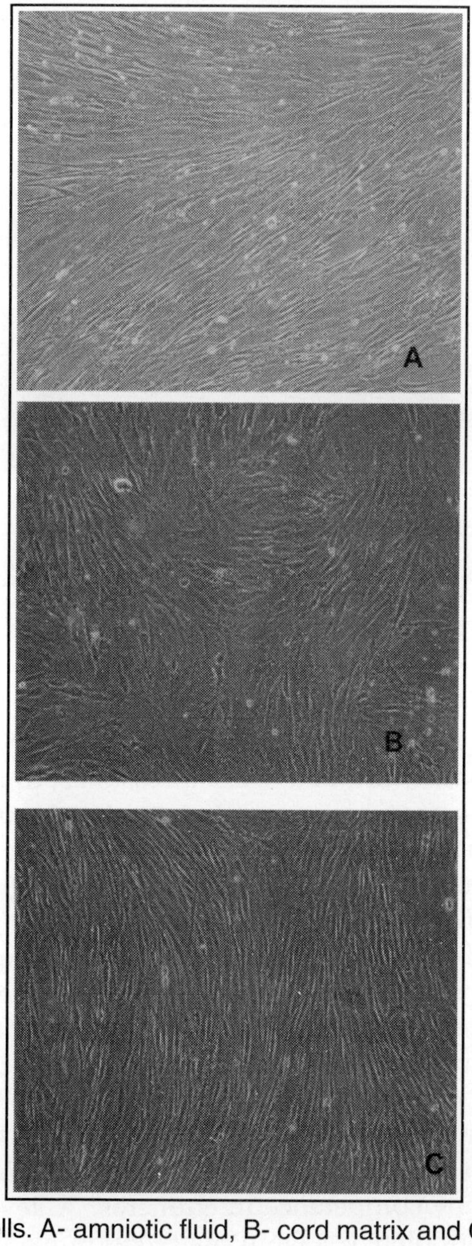

Fig. 4. Buffalo cells. A- amniotic fluid, B- cord matrix and C- fetal fibroblast

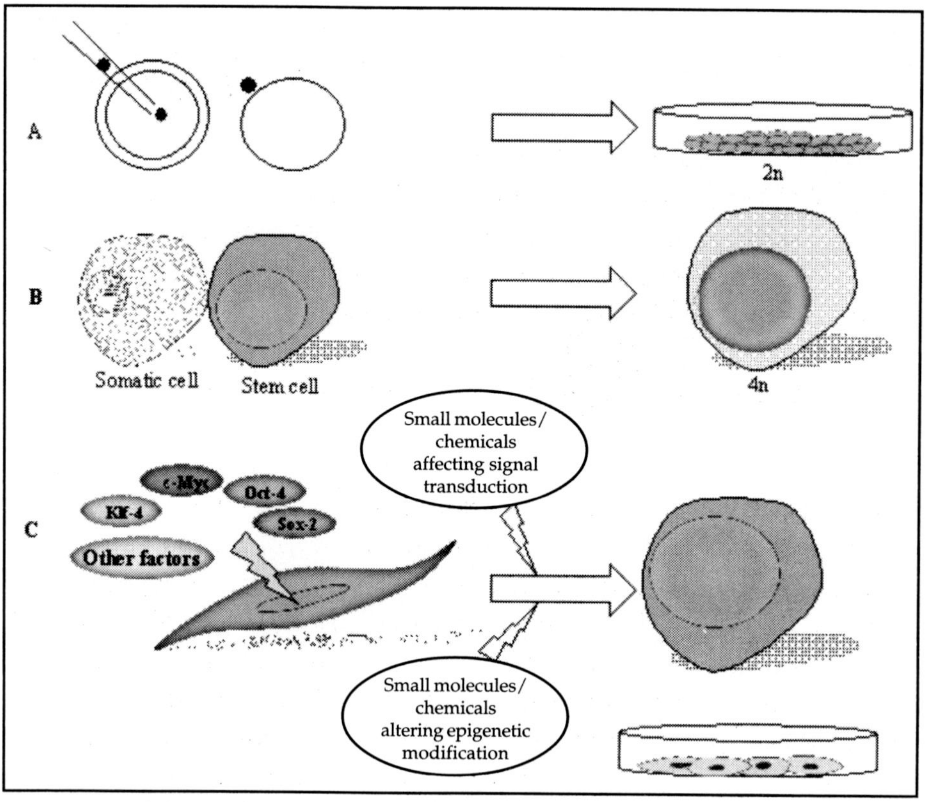

Fig. 5. Scheme for producing pluripotent stem cells. A- by NT, B- by fusion of stem cells with somatic cell and C- iPS cells.

The viral expression vectors used to prepare iPS cells have raised additional concerns. Continued over expression of transcription factors, especially the *c-myc* oncogene may be associated with tumorigenesis. Further, the promoters of these viral vectors appear to be silenced when endogenous gene expression commences during reprogramming (Fenno *et al.*, 2008). This led to the emphasis on use of non-viral methods to induce pluripotency in cells. Among the salient achievements is induction of pluripotency by chemical factors or selected drug-mediated reprogramming of the somatic cells. Isolation of iPSC colonies has been achieved through drug selection mediated activation of the pluripotency-specific gene promoters, Fbx15, Oct4, or Nanog. While Fbx15-iPSCs closely resemble *Oct4*- and *Nanog*-mediated iPS cells in morphology, they are distinct in both epigenetic status and germline competence in chimeras. A key requirement for acquisition of pluripotency is DNA methylation-mediated silencing of exogenous retroviral genes (reviewed in Jaenisch and Young, 2008). Together, review of literature suggests that ES cells and iPS cells are distinguishable by gene expression signatures, and that iPS cells should be considered as unique subtype of pluripotent stem cells.

The information is accumulating on generation and genomics of iPS cells in human and other mammalian species. Perhaps the most striking aspects of these findings include the relatively small complement of factors required to make an Es cell-like cell and the conservation of these factors between species. iPS cells hold great promise for regenerative medicine, because iPS cells circumvent not only immunological rejection but also ethical issues. Since the first report on the derivation of iPS cells in 2006, many laboratories all over the world started research on iPS cells and have made significant progress.

Thus far, reports are lacking on validated ES cell lines from domestic farm animals, such as bovine, swine, and sheep. In this concern, generation of iPS cells from farm animals can be an alternative source of pluripotent cells, and benefit agricultural practice and biomedical research. Recently, porcine iPS cells have been established (Esteban *et al.*, 2009; Wu *et al.*, 2009). Also, no chimera pig or transgenic pig have been produced with pig iPS cells, though mice have been cloned using iPs cells (Boland *et al.*, 2009; Zhao *et al.*, 2009). This has raised the hopes of applying animal iPS cells in gene-targeted or transgenic animal production, as well as to derive iPS cells from other large farm animals. Greater understanding of the molecular mechanisms that mediate reprogramming will ultimately enable the production of safe human iPS cells and multipotential stem cells for use in clinical applications (Tada, 2008).

The pursuit of ES-cell research in animals

The field of adult stem cell research in livestock species is increasing exponentially and encompasses a wide range of topics- from deepening our understanding of cellular development to applying these findings to repair and create organs. Fundamental to the use of any stem cell therapy is a clear understanding of the immunology for allotransplantation as well as autotransplantation for malignancies and immunologically mediated diseases. Scientists and academic centers for stem cell research are using adult stem cells to develop therapies for cancer, stroke, spinal cord injury, autoimmune diseases, and regeneration of bone, cartilage, and other tissues.

There are multiple interests in obtaining pluripotent cells in animals with main expectations in the field of genetic engineering, either to produce cloned animals or transgenic animals **(Table-1)**. Isolation and culture of primary ES cell colonies are the first critical step towards establishment of stable ES cell lines. In fact, the advent of SCNT for reprogramming of the cells has dramatically revolutionized the stem cell technology. Among the livestock species, cattle and pig stem cells are the most widely studied. Authentic bovine ES cell lines have not been established despite progress made for more than two decades. Isolation of pure ICM cells for obtaining the ES cells is one of the major technical constraints in establishing the ES cell cultures in the species, though some innovative approaches like "separate and seed" (Cao *et al.*, 2009) are suggested to improve the technology. The bovine embryonic stem cells lines reported till date vary in morphology and marker expression, such as alkaline phosphatase (AP), stage-specific embryonic antigen-4 (SSEA-4), Nanog, and octamer-binding transcription factor (Oct-4), that normally are associated with the undifferentiated pluripotent state. The bovine ES cells can grow in large, multicellular colonies resembling mouse ES cells and EG cells as well as human EG cells.

Not all of the stem cell research conducted in animals and using animal product is purely anthropocentric; some experiments are aimed at improving the lives of animals. The ES cells offer enormous possibilities of producing large number of offspring by nuclear transfer and performing genetic manipulation for production of cloned transgenic animals. In, addition, the ES cells have applications in basic research, clinical research and in livestock production improvement. The potential improvements in stem cell biology would come in the form of reduction in use of animal for research, better therapeutics for veterinary applications, and conservation of endangered fauna.

Establishing the ES cell lines in farm animals like, buffalo, goat and cattle is of interest for similar reasons to those of mouse and human ES cell lines. This includes basic research such as, comparative embryology and the cell biology of ungulate stem cell maintenance and differentiation. Also, the ES cells may be useful for generating transgenic animals or developing disease-resistant cloned livestock herds, as these cells have a self-renewable capacity more prolonged than normal somatic cell cultures, thus increasing their applications as donor nuclei (Brevini *et al.*, 2008). In addition, the genetic engineering with ES cells may result in synergistic gains in the ability to precisely induce and study genetic alterations in animals.

Stem cells and *in vitro* derivation of gametes

Advances in stem cell research have opened new perspectives for regenerative and reproductive medicine. Stem cells can differentiate under appropriate *in vitro* and *in vivo* conditions into different cell types. *In vitro* derivation of germ cells and viable gametes is a recent emerging field and a promising area of mammalian assisted reproduction. While pluripotency of ES cells is well established, it came as revelation that murine ES cells could develop into primordial germ cells (PGC) *in vitro*, and that extended culture could occasionally form early spermatids or oogonia (Nagano, 2007; Ko and Scholer, 2006; Hubner *et al.*, 2003). Production of gamete precursor cells was achieved initially through allowing the pluripotent stem cell lines to differentiate more or less randomly. The research continues on identifying the precise mechanisms by which differentiation of pluripotent stem cells into gametes can be controlled. Currently, the strategies to create artificial gametes *in vitro* include converting diploid somatic cells from mitotic division to meiotic division directly (somatic cell haploidization), or dedifferentiating somatic cells into pluripotent stem cells (induced pluripotency) and re-differentiating these cells into gametes, or extracting adult stem cells and re-differentiating them into gametes (Nagy *et al.*, 2008). It has been speculated that revolutionary studies showing differentiation of ES cells into germ cells in vitro, and potential oocyte generation *in vivo* from the terminally differentiated cells not only challenge the long-held dogma, but also herald a promising future in the area of mammalian assisted reproduction (Oktem and Oktay, 2008). Childs *et al.* (2008) have highlighted the recent developments in stem cell-derived gametes in humans as well as experimental animal models. While evidences are there that cultured ES cells can produce PGC and germ cells, there remains a question as to whether the germ cells can be derived from adult stem cells residing outside the gonads? Germ cells have been derived from the bone marrow (Nyernia *et al.*, 2006) and fetal skin cells *in vitro* in some experimental models (Dyce *et al.*, 2006).

Because the developmentally competent oocytes are restricted in number and accessibility, a robust system to derive oocytes from stem cells would enable a thorough investigation of the genetic, epigenetic, and environmental factors affecting buffalo oocyte development. Also, the technology may offer ample opportunities for derivation of *in vitro* produced oocytes for procedural refinement of genetic engineering tools in buffaloes, progress in which is hampered severely due to use of abattoir-derived oocytes.

Differentiation of germ cells from stem cells has the potential of becoming a future source of gametes for research use, although further investigation is needed to understand and develop the appropriate niches and culture conditions. Additionally, if genetic and epigenetic methodological limitations could be solved, therapeutic opportunities could be also be considered (Marques-Mari *et al.*, 2009). Concerning livestock applications, maintenance and proliferation of various germline cells derived from ES cells along with establishing protocols for their transplantation may offer new opportunities to treat the reproductive failures in high merit genotypes.

Stem cell research in water buffaloes

Stem cell technology in buffaloes has just begun. Verma *et al.* (2007) have, for the first time, reported ES cell-like cells from inner cell mass (ICM) cells of buffalo embryos produced from IVF of the abattoir-derived oocytes. These cells were found to exhibit ES cell-specific features including formation of embryoid bodies in suspension, expression of elevated levels of AP and Oct-4 (Verma *et al.*, 2007). ES cell lines exhibiting the properties as noted for model ES cell lines were established from parthenogenetically developed buffalo embryos (Sritnaudomchai *et al.*, 2007). The cells termed as PGES (parthenogenetic ES cells) were capable of differentiation into a variety of cell types, and were suggested to be suitable for use as an *in vitro* model to study effects of imprinting on cell differentiation and as an invaluable material for molecular studies (Sritnaudomchai *et al.*, 2007). Huang *et al.*, (2007) have reported on embryonic germ (EG) cells from very early stage buffalo fetuses. These cells were found to be pluripotent, and express ES cells-specific markers including AP, SSEA-1, SSEA-3, SSEA-4 and Oct-4 (Huang *et al.*, 2007, 2008). Adult stem cells were isolated from amnion, amniotic fluid and umbilical cord matrix from buffaloes **(Fig.3)**. These cells have shown characters of stem cells **(Fig.4)**. These cells have been found to survive for more than six months *in vitro*, thus may have potential to be qualified as fetal stem cell lines (Yadav *et al.*, 2008a,b).

A recent report (Xie *et al.*, 2008), suggests that buffalo spermatogonia could differentiate into spermatids *in vitro* as is indicated by analysis of their morphology, PRM2 expression and Raman spectroscopy. However, the normality of the spermatid-like cells was not supported by TP1 expression. However, reports are lacking on the applications of bubaline spermatogonial stem cells. Hopefully, in the next 2-3 years the researchers will come up with experimentally validated role of bubaline spermatogonial stem cells in developing transgenic animals or production of calves derived from the transfer of the embryos produced using spermatogonial stem cells.

Another important area of stem cell research is the transformation of somatic cells into pluripoent stem cells through exogenous introduction of certain factors into the cultured cells. This area of research was initiated in 2006, and has gained enormous momentum. However, till date, the iPS cells among livestock species have been established in pig only (Ezashi *et al.*, 2009).

Opportunities and challenges

For reasons that are unclear the production of ES cells from ICM of blastocysts and the epiblast of slightly older embryos has proven elusive in ungulates (Ezashi *et al.*, 2009). The attempts to establish ES cell lines of domestic animals have been unsuccessful in part due to the inability to develop suitable culture conditions for these species. This is because most of the culture systems for animal stem cells are based on those reported for murine or human ES cells. It is likely that these culture conditions do not support the growth of ES cells in other mammalian species, which differ in factors and cellular signaling cascade indispensable for survival and maintenance of ES cell pluripotency (Renard, 2007). Moreover, lack of cellular as well as molecular markers for validation of pluripotency and species-specificity are yet to be identified. Therefore, if we want to succeed into isolation and establishment of ES cell lines from species other than human and mice, it is imperative that we perform careful cross-species comparison without overlooking species-specific differences (Munoz *et al.*, 2009).

Conclusion

ES cells provide an attractive platform for examining basic developmental processes and disease pathophysiology. A major advantage of embryonic stem cells is that they are capable of differentiating into all types of body cells, as opposed to adult stem cells that seem to be more limited in terms of "to which kind of somatic cells they can be developed." Recently reported improvements in ES cell-related tools have the potential to further accelerate the use of ES cells in this arena. Chief among these advances are the novel tools for genetic manipulation of ES cells and improvements in ES cell culture conditions. In view of the ethical problems in deriving ES cells embryonic stem cells, the researchers are interested in discovering ways of obtaining embryonic stem cells or pluripotent stem cells without compromising the embryo from which the cells are drawn. In domestic ungulates, the embryonic stem cells, embryonic germ cells and adult stem cells are still at initial stages of development.

Table 1. Prospects of stem cell technology in livestock

1. Improvement in male fertility through spermatogonial stem cell transplantation.
2. Development of stem cell-based novel therapies to treat animal specific diseases.
3. Improving cloning efficieny through use of stem cells as donor nuclei
4. Genetic manipulation in stem cells for production of transgenic animals
5. Using spermatogonial stem cells for production of transgenic animals
6. Studying the basic cellular and molecular aspcets of buffalo embryonic development
7. *In vitro* meat production from the muscle and adipose stem cells.
8. To resolve problems related to the livestock therapies e.g. regeneration of teat tissue damaged by mastitis in high milk-yielding animals or regeneration of damaged ligaments
9. Application of pluripotent stem cells in veterinary and human drug testing and screening.
10. The use of stem cell technology in farm animals may overcome current limitations in efficient gene transfer by providing an abundance of stem cells to be genetically manipulated by using conventional recombinant DNA techniques

References

Boland MJ, Hazen JL, Nazor KL, Rodriguez AR, Gifford W, Martin G, Kupriyanov S and Baldwin KK. **2009**. Adult mice generated from induced pluripotent stem cells. *Nature*. **461**, 91-94.

Brevini TA, Antonini S, Pennarossa G and Gandolfi F. **2008**. Recent progress in embryonic stem cell research in domestic species. Review. *Reprod. Domest. Anim*. **43 (Suppl.)**, 193-199.

Brons IG, Smithers LE, Trotter MW, Rugg-Gunn P, Sun B, Chuva de Susa Lopes SM, Howlett SK, Clarkson A, Ahrlund-Richter L, Pedersen RA, Vallier L. **2007**. Derivation of pluripotent epiblast stem cells from mammalian embryos. *Nature*. **448**, 191-195.

Cao S, Wang F, Chen Z, Liu Z, Mei C, Wu H, Huang J, Li C, Zhou L and Liu L. **2009**. Isolation and culture of primary bovine embryonic stem cell colonies by novel method. *J. Exp. Zool. A Ecol. Genet. Physiol.* **311**, 368-376

Childs AJ, Saunders PT, Anderson RA. **2008**. Modelling germ cell development in vitro. *Mol. Hum. Reprod.* **14**, 501-511.

Chuang CK, Sung LY, Hwang SM, Lo WH, Ghen HC, Hu YC. **2007**. Baculovirus as a new gene delivery vector for stem cell engineering and bone tissue engineering. *Gene Ther.* **14**, 1417-1424.

Dyce PW, Wen L, Li J. **2006**. *In vitro* germline potential of stem cells derived from foetal porcine skin. *Nature Cell Biol.* **8**, 384-390.

Esteban MA, Xu J, Yang J, Peng M, Qin D, Li W, Jiang Z, Chen J, Deng K, Zhong M, Cai J, Lai L and Pei D. **2009**. Generation of induced pluripotent stem cell lines from Tibetan miniature pig. *J. Biol. Chem.* **284**, 17634-17640.

Evans MJ and Kaufman MH. **1981**. Establishment in culture of pluripotential cells from mouse embryos. *Nature*. **292**, 154–156.

Ezashi T, Telugu BP, Alexenko AP, Sachdev S, Sinha S and Roberts RM. **2009**. Derivation of induced pluripotent stem cells from the pig somatic cells. *Proc. Natl. Acad. Sci. USA*. **106**, 10993-10998.

Fenno LE, Ptaszek LM and Cowan CA. **2008**. Human embryonic stem cells: emerging technologies and practical applications. *Curr. Opin. Genet. Dev.* **18**, 324-329.

Martin G. **1981**. Isolation of a pluripotent cell line from early mouse embryos cultured in medium conditioned by teratocarcinoma stem cells. *Proc. Natl. Acad. Sci. USA*. **78**, 7634–7638.

Galli C and Lazarri G. **2008**. The manipulation of gametes and embryos in farm animals. *Reprod. Domest. Anim.* **43(Suppl. 2)**, 1-7.

Huang B, Xie TS, Shi, DS, Li T, Wang XL, Mo Y, Wang ZQ, Li MM. **2007**. Isolation and characterization of EG-like cells from Chinese swamp buffalo (*Bubalus bubalis*). *Cell Biol. Int.* **31**, 1079-1088.

Huang B, Li T, Wang XL, Xie TS, Lu YQ, da Silva FM, Shi DS. **2008**. Generation and characterization of embryonic stem-like cell lines derived from *in vitro* fertilization buffalo (*Bubalus bubalis*) embryos. *Reprod. Domest. Anim.* (in press).

Hubner K, Fuhrmann G, Christenson LK, Kehler J, Reinbold R, De La Fuente R, Wood J, Strauss JF 3rd, Boiani M. **2003**. Derivation of oocytes from mouse embryonic stem cells. *Science*. **300**, 1251-1256.

Jaenisch R and Young R. **2008**. Stem cells, the molecular circuitry of pluripotency and nuclear reprogramming. *Cell*. **132**, 567-582.

Ko K and Scholer HR. **2006**. Embryonic stem cells as a potential source of gametes. *Semin. Reprod. Med.* **24**, 322-329.

Lawson KA and Pederson RA. **1992**. Clonal analysis of fate during gsatrulation and early neurulation in the mouse. **In**: CIBA Foundation Symposium, 165 New York, John Wileys and Sons, USA, pp3-26.

Lo WH, Hwang SM, Chuang CK, Chen CY, Hu CY. **2009**. Development of hybrid baculoviral vector for sustained transgene expression. *Mol. Ther.* **17**, 658-666.

Lufino MM, Esder PA and Wade-Martins R. **2008**. Advances in high-capacity extrachromosomal vector technology: episomal maintenance, vector delivery, and transgene expression. *Mol. Ther.* **16**, 1525-1538.

Marques-Mari AI, Lacham-Kaplan O, Medrano JV, Pellicer A, Simón C. **2009**. Differentiation of germ cells and gametes from stem cells. *Hum. Reprod. Update*. **15**, 379-390

Martin GR. **1981**. Isolation of pluripotent cell line from early mouse embryos cultured in medium conditioned by teratocarcinoma stem cells. *Proc. Natl. Acad. Sci. USA*. **78**, 7634-7638.

Munoz M, Rodriguez A, De Frutos C, Caamano JN, Diez C, Facal N, Gomez E. **2008**. Conventional pluripotency markers are unspecific for bovine embryonic-derived cell-lines. *Theriogenology*. **69**, 1159-1164.

Munoz M, Trigal B, Molina I, Diez C, Cammano JN and Gomez E. **2009**. Constraints to progress in embryonic stem cells from domestic species. *Stem Cell Rev. Rep.* **5**, 6-9.

Nagano MC. **2007**. In vitro gamete derivation from pluripotent stem cells: progress and perspective. *Biol. Reprod.* **76**, 546-551.

Nagy ZP and Kerkis I and Chang CC. **2008**.Development of artificial gametes. *Reprod. Biomed. Online*. **16**, 539-544.

Nayernia K, Lee JH, Drusenheimer N, Nolte J, Wulf G, Dressel R, Gromoll J, Engel W. **2006**. Derivation of male germ cells from bone marrow stem cells. *Lab. Invest.* **86**, 654-663.

Oktem O and Oktay K. **2008**. Stem cells: a perspective on oocytes. *Ann. NY Acad. Sci.* **1127**, 20-26.

Renard JP, Maruotti J, Jouneau A and Vignon X. **2007**. Nuclear reprogramming and pluripotency of embryonic cells: application to the isolation of embryonic stem cells in farm animals. *Theriogenology*. **68 (Suppl. 1)**, S196-S205.

Robertson EJ. **1987**. Embryo-derived stem cell lines. **In**: *Teratocarcinomas and embryonic stem cells: a practical approach*. E. J. Robertson (Ed.), (pp. 71–112). Oxford: IRL.

Sritanaudomchai H, Pavasuthipaisit K, Kitiyanant Y, Kupradinun P, Mitalipov S and Kusamran T. **2007**. Characterization and multilineage differentiation of embryonic stem cells derived from a buffalo parthenogenetic embryo. *Mol. Reprod. Dev.* **74**, 1295–1302.

Tada, T. **2008**. Genetic modification-free reprogramming to induced pluripotent cells: fantasy or reality? *Cell Stem Cell.* **3**, 121-122.

Takahashi K and Yamanaka S. **2006**. Induction of pluripotent stem cells from mouse embryonic and adult fibroblast cultures by defined factors. *Cell.* **126**, 663–676.

Talbot NC and Blomberg A. **2008**. The pursuit of ES cell lines of domestic ungulates. *Stem Cell Rev.* **4**, 235-254.

Tesar PJ, Chenoweth JG, Brook FA, Davies TJ, Evans EP, Mack DL, Gardner RL and McKay RD. **2007**. New cell lines from mouse epiblast share defining features with human embryonic stem cells. *Nature.* **448**, 196-199.

Thomson JA, Itskovitz-Eldor J, Shapiro SS, Waknitz MA, Swiergiel JJ, Marshall VS, Jones JM. **1998**. Embryonic stem cell lines derived from human blastocysts. *Science.* **282**, 1145–1147. (Erratum in: *Science* **1998** Dec 4; **282** (5395), 1827).

Thomson JA and Marshall VS. **1998**. Primate embryonic stem cells. **In**: *Current Topics in Developmental Biology.* Pedersen RA and Schatten GP (Eds.), San Diego, CA: Academic. (pp. 133–165).

Verma V, Gautam SK, Singh B, Manik RS, Palta P, Singla SK, Gosawami SL and Chauhan MS. **2007**. Isolation and characterization of embryonic stem cell-like cells from in vitro-produced buffalo (*Bubalus bubalis*) embryos. *Mol. Reprod. Dev.* **74**, 520-529.

Vogel G. **2007**. A knockout award in medicine. *Science.* **318**, 178-179.

Wang Z, Ruan J and Cui D. **2009**. Advances and prospect of nanotechnology in stem cells. *Nanoscale Res. Lett.* **4**, 593-605.

Wu Z, Chen J, Ren J, Bao L, Liao J, Cui C, Rao L, Li H, Gu Y, Dai H, Zhu H, Teng X, Cheng L and Xiao L. **2009.** Generation of pig-induced pluripotent stem cells with a drug-inducible system. *J. Mol. Cell. Biol.* **1**, 46-54.

Xie B, Qin Z, Huang B, Xie T, Yao H, Wei Y, Yang X, Shi D, Jiang H. **2008**. *In vitro* culture and differentiation of buffalo (*Bubalus bubalis*) spermatogonia. *Reprod. Domest. Anim.* (in press)

Yadav PS, Tokas J, Sharma RK, Singh I and Sethi RK. **2008a**. Buffalo amniotic fluid, umbilical cord matrix and early fetal explants as possible source of adult stem cells. **In**: *IX Annual Conference of Indian Society of Animal Genetics and Breeding*, Held at NASC Complex Delhi July 3-4, 2008.

Yadav RP, Yadav PS, Nanda T and Singh I. **2008b**. Isolation and culture of stem cells like cells from buffalo amnion. **In**: *First International Stem Cell Submit-2008*, Held at IIT, Chennai, Nov 14-16 2008.

Zhao XY, Li W, Lv Z, Liu L, Tong M, Hai T, Hao J, Guo CL, Ma QW, Wang L, Zeng F and Zhou Q. **2009**. iPS cells produce viable mice through tetraploid complementation. *Nature*. **461**, 86-90.

—□□□—

9

Biopharming Using Livestock

"Large-scale production of biopharmaceuticals by conventional bioreactor techniques is limited by low transgenic efficiency and low expression of foreign proteins in vitro. With advances in the efficiency of creating transgenic farm animals, it is now possible to produce, isolate and purify larger quantities of therapeutic proteins and nutraceuticals, and high quality food products at comparatively reasonable costs".

Introduction

Human community has extensively used the plant extracts or microbial products for the centuries as pharmaceuticals. In an effort to overcome the problems associated with inadequate production of therapeutically important biomolecules, e.g. human immunoglobulin, various technologies have been developed. For example, the production of human immunoglobulin by recombinant methods in mammalian cell and tissue culture (Table-1) is a routine practice. Particularly, expression of recombinant human immunoglobulin in chinese hamster ovary CHO expression systems is well known and is currently utilized for the production of several biomolecules and chimeric antibodies. The eukaryotic cells are used to synthesize larger size and complex proteins compared to the prokaryotes which are employed to produce smaller and simpler proteins. The prokaryotes, as host for production of eukaryotic proteins have certain obvious limitations, e.g. they can't synthesize complex proteins (like monoclonal antibodies or blood coagulation factors) which must be matured by post-translational modifications (glycosylation, carboxylation, phosphorylation, and sulphatation etc.) to be active or stable *in vivo*. The baculovirus as expression system is used for production of eukaryotic recombinant proteins due to its ability to carry out more complex post-translational modifications and ease of purification. Alternatively, yeasts (*Saccharomyces cerevisiae* and *Pichia pastoris*), insects, animal cells and transgenic plants (e.g. *Arabidopsis thaliana*) are used for bulk production of therapeutic proteins. Indeed, these cells are eukaryotes and they may potentially synthesize and secrete large amounts of post-translationally modified or matured proteins. However, due to certain limitations of these hosts, e.g. inability of yeasts to add multiple disulphide (S-S) bridges into the proteins, inefficient glycosylation of the human recombinant proteins by insect cells make them less desirable hosts. Further, the major drawback of mammalian cell culture system is their inability to produce higher quantity of proteins (Nieman and

Kues, 2007; Houdbine, 2009). The limited lifespan of normal mammalian cells *in vitro*, and the problems associated with the acquisition of materials from normal and pathological tissues, make them less suitable as producers of recombinant proteins.

The use of animal proteins as pharmaceuticals began in the 1920s with the extraction and recovery of insulin from pig pancreas. Much later, in the early 1980s, larger quantities of human insulin were prepared using recombinant DNA technology. At present, numerous other proteins and pharmaceuticals including human growth hormone are being produced using this technique for humans as well as livestock therapeutic applications (Houdbine, 2009). Preferably, the transgenic ungulate expresses an immunoglobulin chain or antibody from another genus, such as an antibody from any other mammal. Transgenic animals like goats, mice, cows, pigs, rabbit, and sheep are being developed as exogenous-proteins production systems; some aquatic animals are also being utilized. Particularly preferred ungulates are pigs, sheep, goats, buffaloes, antelopes, oxen, horses, donkeys, mule, deer, elk, caribou, water buffalo, camels, llama, etc. Not exclusively, but the animals are being used as models for novel surgical strategies, testing of biodegradable implants and sources of tissue replacements, such as skin and heart valves is also being tested in these animals. Domestic animals are harnessing as bioreactor for production of biopharmaceuticals through gene-farming with efficiency far greater than any conventional microbial or cell-culture production system.

Biopharming through animal transgenesis

Biopharming is a new generation technology in which transgenic organisms are produced for bulk production of pharmaceuticals. If animals are tailored genetically to synthesize the recombinant proteins, the process is called '*animal pharming*'. Genetic engineering involves the techniques and protocols that result in a directed alteration in the genotype of an organism. Whereas the advanced molecular genetics, recombinant DNA technology and reproductive biology enable the generation of transgenic animals, the modern technologies such as stem cell technology and nanotechnology have additionally raised the hopes for a revolution in the field of animal transgenesis. For animals, the term "transgenic" originally refers to the stable introduction of exogenous genetic material into their germ line. This definition of transgenic animals has been extended for two reasons. First, the further developments in animal genetic engineering enables not only the additive gene-transfer (gain of function), but also developing the animals with deleted gene-transfer (knock-out, loss of function), replacement gene-transfer (knock-in, exchange of gene function) and spatial-temporal gene-transfer (conditional knock-out) (Clark and Whitelaw, 2003; Ristevski, 2005). Second, in addition to germline integration of transgenes, somatic gene-transfer approaches may result in (mostly transient) gene-expression with longest duration being a lifespan and no transmission of the transgenes to the progeny. Although somatic gene-transfer has potential for livestock production purposes, the livestock transgenesis is more beneficial for the development of DNA-based vaccines (Dunham, 2002) or nutraceuticals.

Transgenic animals may offer particular attractive possibilities **(Fig.1)** to prepare biologically stable recombinant pharmaceuticals. The advantages are high and low cost production as well as high quality of the proteins. Initially, the transgenic technology was developed as research tool to study gene function in mice. While the mouse is an ideal scientific model organism for mammalian species, the transgenic technology had immediate applications for both agriculture and biomedical purposes when extended to livestock species (Laible, 2009). The development of transgenic animal was initiated as early as 1980s, (Table-1) with the development of first transgenic mouse using pronuclear microinjection of the exogenous DNA. Hammer *et al.* (1985) for the first time, reported the first births of transgenic rabbits, sheep and pigs using microinjection of foreign DNA into the pronuclei of the zygotes of these species. However, the technique could not be exploited commercially due to certain problems including random insertion of the exogenous genes into the target genome, uncontrolled copy number of the genes incorporated and the reduced survival of the embryos injected with foreign genes. Another problem was the expression of recombinant proteins in host. This is because blood at most of the time cannot store high levels of recombinant proteins which are naturally too unstable. Moreover, biologically active proteins in blood may possibly alter the physiology or overall health status of the host.

Methods of producing transgenic animals

Various methods (Fig.2) **(Table-2)** are in use to transfer exogenous genes in target animal species. Development of an efficient delivery system is important for basic research of cell biology and therapeutic applications. Use of a particular technique to produce a transgenic animal depends much on the target species. The main routes to transgensis in mammalian species include:

1. Microinjection of exogenous DNA into the pronucleus of a fertilized oocyte (zygote) or into the somatic cells or pluripotent stem cells
2. Generating the lentiviral vectors or transposons and integration of viral vector carrying gene of interest into an oocyte or early embryo
3. Incorporation of genetically engineered pluripotent stem cell into early embryo for generating chimeric animals harboring normal and transformed cells
4. Using the genetically-altered somatic cell or ES cell donor nucleus for producing cloned transgenic animal; and
5. Sperm-mediated gene transfer

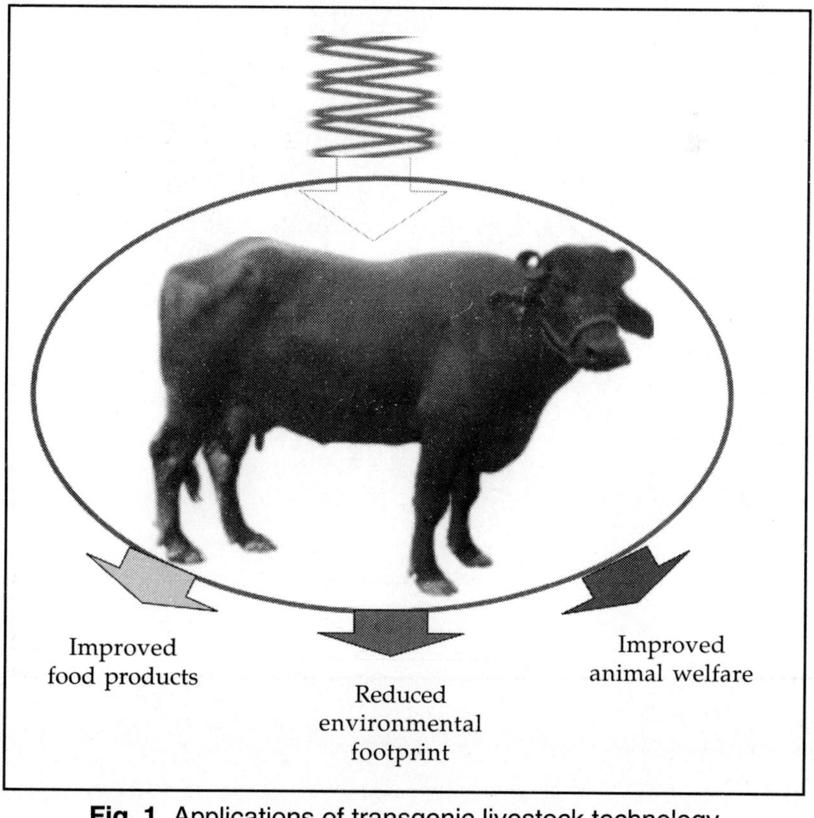

Fig. 1. Applications of transgenic livestock technology

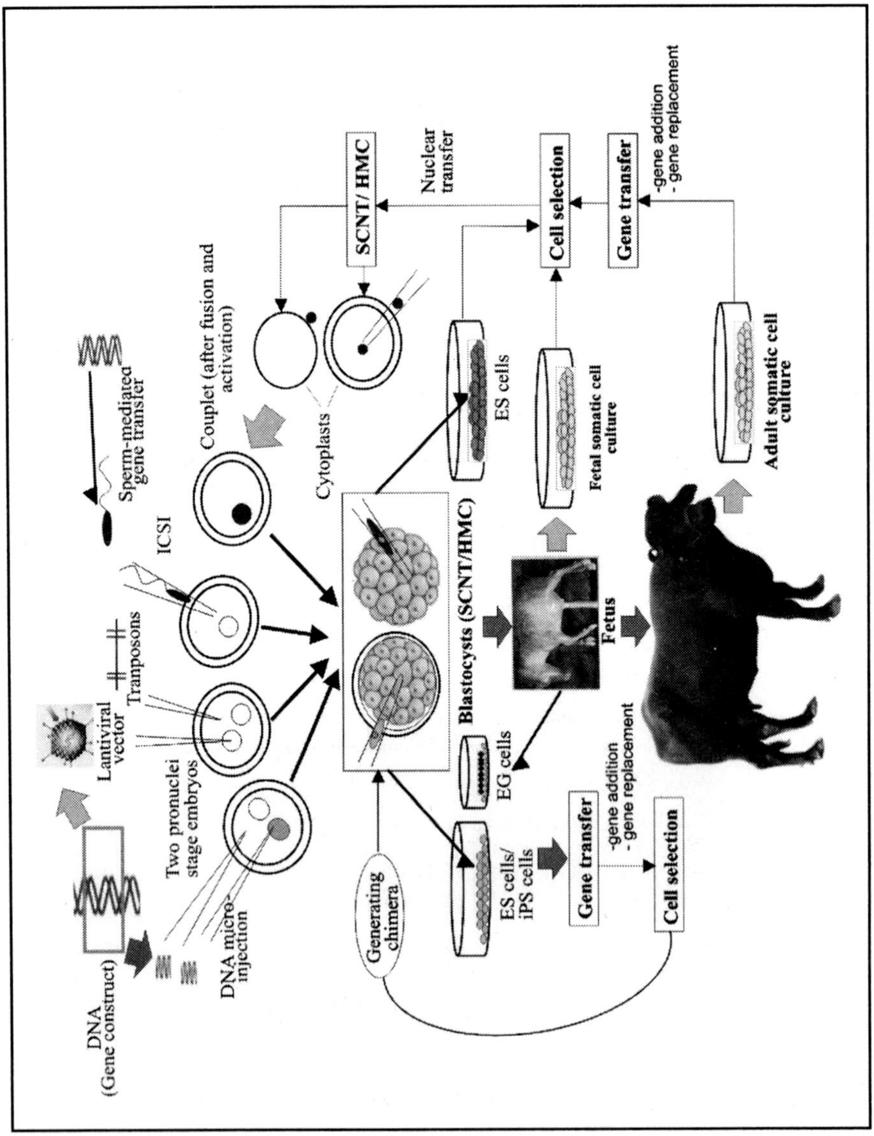

Fig.2. Scheme for producing transgenic animal

Among the various methods of producing transgenic animals, the microinjection to single cells is more widely used in the studies of transduction-challenged cells to mechanically transfer the DNAs, RNAi, sperm, peptides and drugs (Zhang and Lu, 2008).

Mammary gland as bioreactor

The mammary gland is the preferred production site mainly because of the quantities of proteins that can be produced by employing various mammary gland-specific promoters and the ease of production and subsequent *in vitro* purification of the therapeutic products. Numerous proteins have been produced in large amounts in the mammary glands of transgenic sheep, goat, cattle, pig and rabbit (Kues and Niemann, 2004).

In theory, large quantities of the human protein can be produced in the animal's milk and subsequently purified for therapeutic applications. An early example of this technology (Table-3) by John Clark and colleagues was the production of transgenic sheep expressing the human blood-clotting factor IX which is required by patients with hemophilia. The human factor IX gene was expressed under the control of sheep genome that normally turns on the β-lactoglobulin gene in the mammary tissue. Blood, milk, (Houdbine, 2000, 2002), egg white (Zhu *et al.*, 2005; Lillico *et al.*, 2007; Yang *et al.*, 2008), urine, and seminal plasma, silk gland (Royer *et al.*, 2005) and insect larvae haemolymph (Markaki *et al.*, 2007) are other possible systems. Milk is presently the most mature system to produce recombinant proteins from transgenic animals.

Combined with tissue-specific functional promoter elements, non-mammalian genes can also be expressed in farm animals to modify intermediary metabolism. With advances in the efficiency of creating, transferring and expressing the foreign genes in transgenic farm animals, it is now possible to isolate and purify the larger quantities of therapeutic proteins at comparatively reasonable costs.

Buffalo as possible species for biopharming

The animals with higher milk yield, higher fertility, non transmission of human diseases, and resistance to prion diseases could be the choice to produce recombinant proteins. As the buffaloes are known to yield high protein milk and, therefore, they may possibly have better potential as live bioreactor for producing therapeutically important biomolecules in the milk. While initially, the concern of buffalo transgenesis may be the '*biopharming*', larger-scale applications could include production of nutriceuticals or orally taken therapeutics. Further applications may include production of specific antibody titers in milk to confer immunity to neonatal calves against deadly infections. To develop transgenic buffalo embryos a pronuclear microinjection method has been developed recently. Gene construct for synthesis of green fluorescent protein (GFP) was incorporated into the early stage zygotes through a pro-nuclear microinjection and the transgenic embryos expressing GFP were found to grow to blastocyst stages (Verma *et al.*, 2008). In another study, Huang *et al.*, (2008) have reported that gene-transfected EG-like cells have the ability to form chimeric embryos after injecting them into early stage-embryos, and undergo reprogramming after standard nuclear transfer, and the process can be used to produce cloned transgenic buffaloes through embryo chimera or standard SCNT techniques.

However, several biological barriers present a challenge for gene-delivery into the cell, as a result, the integration of the exogenous genetic material remains one of the major challenges in producing transgenic stem cells or embryos. On the other hand, developments in nanotechnology now offer numerous non-viral vectors that have been fabricated and found capable of transmitting the genes into a cell and even into specific subcellular compartments like mitochondria. In recent years stem cells and nanotechnology are gradually intercrossing to form a new emerging field, that is, nanotechnology in stem cells or stem cell nanotechnology (Wang *et al.*, 2009). The studies of interaction between nanomaterials or nanostructures and stem cells have made great advances (Wang *et al.*, (2009). Moreover, the innovations in mechanisms of intracellular delivery and integration of foreign genes into the genome through highly efficient, safe and suitable vectors like baculovirus vectors (Chuang *et al.*, 2007; Luffino *et al.*, 2008; Lo *et al.*, 2009) or multifunctional nanoparticles, liposomes or biodegradable polymeric vectors may hopefully result in development of novel stem cell technologies which may advance the existing concept of '*biopharming*' by using buffalo mammary gland as live bioreactors. The genetic engineering tools along with advances in cell culture technology may particularly be useful when used in conjunction, and the achievement of the goal seems closer in future.

Opportunities and challenges

Genetic modification of large animals is a difficult task and remains very much in the experimental stages. It is difficult to foresee production of transgenic livestock contributing to improved efficiency of animal production until the efficiency of the process of producing transgenics is dramatically improved. Owing to integration and expression of transgenes, insertional mutagenesis in host tissues and pathologies, the side effects of the technology cannot be ruled out, which raise the concern about the health of transgenic animals. Further systematic studies on the effect of genetic engineering on health of the animals, and the recombinant products on humans are scarce. Research programs on transferring prokaryotic genes to ruminants to introduce cysteine, threonine, and lysine biosynthesis or a functional glyoxylate cycle (for conversion of the major rumen metabolite acetate to glucose) have so far failed. Emerging new technologies, however, may soon revolutionize the scope and efficiency of genetic modification of livestock, which in turn could allow application of transgenic technologies to modify agriculturally significant traits of livestock species. Researchers have initiated to produce genetically modified buffaloes embryos. For application of genetic engineering to successfully exploit the buffalo as live bioreactor, extensive research is needed on preparing the suitable gene-construct, efficient transfer of the genes into cell or embryo, controlled expression of the exogenous or recombinant proteins in target tissues, and improving development of the cloned embryos.

As with the case of genetically modified (GM) crops, future developments are likely to be highly contingent on public acceptability of GM or cloned livestock. Moreover, the guidance from the U.S. Food and Drug Administration on the commercialization of genetically engineered animals has sparked renewed discussion over the ethical, consumer, and regulatory implications of transgenesis in animal agriculture (Greger, 2009). Unless breeding goals are redefined to reflect social concerns, the occurrence and magnitude of undesirable side effects may increase and consumer confidence in the nascent technology will remain undermined.

Conclusion

The generation of GM animals or plants with gene-targeted deletions or modifications is a powerful tool to analyze gene function, study disease and produce genetically-altered organisms of economic interest. Because many different steps are involved, the success rates are often low- usually one or two per cent. This imposes an enormous cost on creation of genetically tailored large ruminants like cattle and buffalo; as such most work has been done in mice, pigs and sheep. Genetic engineering tools for production of therapeutic molecules in milk, augmentation of quality milk and meat production in farm animals as well as for creating models of human diseases is increasing. Further, the need for transgenesis for facilitating the xenotransplantation from animals to humans is realized globally. Use of buffaloes as transgenic animals for production of biomolecules for human medical use has enormous potential, and should be exploited on priority. Genetic engineering along with recent advances in genomics to produce transgenic animals is likely to prove practically viable.

Table 1. Animal cell-lines used for production of therapeutic proteins*

Cell type	Animal source
Mammalian	
Myeloid cell lines	
NS0	Mouse
SP2=0.Ag8	Mouse
63Ag8.653	Mouse
YO	Rat
YB20	Rat
Non-myeloid cell lines	
CHO-K1	Hamster
CHO-I.761 h	Hamster
BKH	Hamster
CV-1	Monkey
COS	Monkey
Amphibian	
X. laevis	Frog
Insect	
S. frugiperda	Insect

(* From Redwan, 2009) CHO :

Table 2. A summary of various tools and methods used in producing transgenic animals, their salient features and major limitations.

Methods	Salient features	Major limitations
Plasmids/ cosmids	Easy to obtain and prepare, excellent safety profiles	Low gene carrying capacity
Liposomes	Long term gene-expression, up to several months depending on context	Low efficiency, but variable depending on cell type
Electroporation	Remains the mainstay of production of gene-targeted applications	May cause transient but reversible disruption of cell membrane / cell lines
Viral-mediated gene-transfer	Ease of transfection of the the cell or zygote; larger capacity of carrying the insert (e.g., lantiviruses)	Hazardous work, threat of escape of viruses from the laboratories, and integration of viral genes into the human or animal genome
	Ease of injection or infection of oocytes or embryos	With the exception of adeno-associated viruses, viral methods are generally incompatible with gene-targeting by homologous recombination
Sperm-mediated DNA transfer	Easy to use (IVF or ICSI)	Low efficiency, low fertilizing capacity of the sperm carrying exogenous genes
Microinjection	Ease of injection (Zhang and Lu, 2008)	Random integration of the gene into host's genome, low efficiency, frequent incedences of mosaicism, poorly efficeint in ruminants
SCNT	Production of complete transgenic organism	The low efficiency of the technique; lack of precise methods to introduce exogenpus genes into the genome of cells to be used as donor nuclei

Nanotechnology	Efficient delivery of the genes, dissolution of the the vector after delivery of genes into host/ cell Improved cell culture technology permits transfer of larger gene constructs	The technique is still in intial phases of validation due to lack of data

Table 3. Milestone events in mammalian transgenesis

Year	Achievement	Strategy and outcomes (reference)
1980	First transgenic mice	Microinjection of purified DNA (Gordon *et al.*, 1980)
1982	Transgenic mice	Effect of growth hormone elucidated using transgenic technology (Palmiter *et al.*,1982)
1985	First transgenic sheep and pigs	Microinjection of exogenous genes into the pronucelus of a zygote (Hammer *et al.*, 1985)
1986	Embryonic cloning of sheep	NT in sheep using ES cells as donor nuclei (Willadsen, 1986)
1987	Transgenic mice	Expression of sheep β-lactoglobulin in the milk of transgenic mice (Simons *et al.*, 1987)
1989	First animal bioreactor	Bulk production of human factor IX (Clark *et al.*, 1989)
1989	Transgeic pigs	Pigs expressing growth hormone (Pursel *et al.*, 1989)
1991	Therapeutic proteins	High-level expression of active human α-1- antitrypsin in milk (Wright *et al.*, 1991)
1992	Disease-resistant pigs	First attempt at producing influenza resistant transgenic pigs (Muller *et al.*, 1992)
1996	Transgenic sheep	Improving wool production through transgenic animals (Demak *et al.*, 1996)
1997	Birth of 'Dolly'	Reprogramming of adult somatic cell was proved for the first time in mammals (Wilmut *et al.*, 1997)

1997	Transgenic sheep	Random integration of the gene construct for human factor IX in transgenic sheep produced by NT (Schniek *et al.*, 1997)
1998	Transgenic cattle from fetal fibroblasts and NT	Random integration of the gene construct (Cibelli *et al.*, 1998)
1998	Transgenic cattle	Birth of transgenic calf by transfection of oocytes with helper viruses (Chan *et al.*, 1998)
2000	Gene targeting in sheep	First transgenic livestock species produced thorugh gene-replacement and NT (McCrath *et al.*, 2000)
2001	Transgenic pigs	Production of transgenic pig with ecofriendly exogenous genes (phytase) adressing envinonmental issues (Golovan *et al.*, 2001)
2002	Trans-chromosomal cattle	Use of artificial chromosome to produce transgenic cattle (Kuroiva *et al.*, 2002)
2002	Knockout in pigs	Heterozygous knockout pigs were produced (Dai *et al.*, 2002; Lai *et al.*, 2002)
2003	Homozygous knockout	Homozygous knockout pigs were produced (Phelps *et al.*, 2003)
2003	Transgenic pigs	Development of strategy for gene transfer into mammalian zygotes via lentiviruses (Hofman *et al.*, 2003)
2006	Transgenic goats	Production, processing and purification of recombinant product i.e. human lysozyme in the milk of transgenic goats (Maga *et al.*, 2006)
2006	Conditional transgene expression	Pronucleus DNA injection in pigs and crossbreeding (Kues *et al.*, 2006)
2009	Transgenic buffalo embryos production	Effects of various culture media and supplements on production of cloned transgenic embryos (Huang *et al.*, 2008; Verma *et al.*, 2008; Wadhwa *et al.*, 2009;) were elucidated
2009	Recombinant drug for use	First FDA approval of transgenic animal drug for human applications (Kling, 2009)

References

Chan AW, Homan EJ, Ballou LU, Burns JC and Bremel RD. **1998.** Transgenic cattle produced by reverse-transcribed gene transfer in oocytes. *Proc. Natl. Acad. Sci. USA.* **95,** 14028–14033.

Chuang CK, Sung LY, Hwang SM, Lo WH, Ghen HC, Hu YC. **2007**. Baculovirus as a new gene delivery vector for stem cell engineering and bone tissue engineering. *Gene Ther.* **14,** 1417-1424.

Cibelli JB, Stice SL, Golueke PL, Kane JJ, Jerry J, Blackwell C, Ponce de Leon FA and Robl JM. **1998.** Transgenic bovine chimeric offspring produced from somatic cell-derived stem like cells. *Nat. Biotechnol.* **16,** 642–646

Clark AJ and Whitelaw B. **2003**. A future for transgenic livestock. *Nat. Rev. Genet.* 2003. **4**, 825-833.

Clark AJ, Bessos H, Bishop JO, Brown P, Harris S, Lathe R, McClenaghan M, Prowse C, Simons JP, Whitelaw CB and Wilmut I. **1989**. Expression of human anti-hemophilic factor IX in the milk of transgenic sheep. *Biotechnology.* **7,** 487-492.

Dai Y, Vaught TD, Boone J, Chen SH, Phelps CJ, Ball S, Monahan JA, Jobst PM, McCreath KJ, Lamborn AE, Cowell-Lucero JL, Wells KD, Colman A, Polejaeva IA, Ayares DL. **2002**. Targeted disruption of the alpha1, 3-galactosyltransferase gene in cloned pigs. *Nat. Biotechnol.* **20,** 251–255.

Demak S, Jay NP, Barrell GK And Bullock DW. **1996.** Targeting gene expression to the wool follicle in transgenic sheep. *Biotechnology (NY).* **14,** 181-184.

Dunham SP. **2002**. The application of nucleic acid vaccines in veterinary medicine. *Res. Vet. Sci.* **73,** 9-16.

Golovan SP, Meidinger RG, Ajakaiye A, Cottrill, M, Weiderkehr MZ, Barney DJ, Plante C, Pollard JW, Fan MZ, Hayes MA, Laursen J, Hjorth, JP, Hacker RR, Phillips JP and Forsberg CW. **2001**. Pigs expressing salivary phytase produce low-phosphorus manure. *Nat. Biotechnol.* **19**. 741-745. Erratum in: *Nature Biotechnol.* **2001. 19,** 979.

Gordon JW, Scangos GA, Plotkin DJ, Barbosa JA, Ruddle FH. **1980.** Genetic transformation of mouse embryos by microinjection of purified DNA. *Proc. Natl. Acad. Sci. USA.* **77,** 7380-7384.

Greger M. **2009**. Trait selection and welfare of genetically engineered animals in agriculture. *J. Anim. Sci.* (*in press*)

Hammer RE, Pursel VG, Rexroad CE Jr., Wall RJ, Bolt DJ, Ebert KM, Palmiter RD and Brinster RL. **1985**. Production of transgenic rabbits, sheep and pigs by microinjection. *Nature.* **315,** 680–683.

Hofmann A, Kessler B, Ewerling S, Weppert M, Vogg B, Ludwig H, Stojkovic M, Boelhauve M, Brem G, Wolf E, Pfeifer A. **2003**. Efficient transgenesis in farm animals by lentiviral vectors. *EMBO Rep.* **4**, 1054–1060.

Houdbine LM. **2000**. Transgenic animal bioreactors. *Transgenic Res.* **9**, 305-320.

Houdbile LM. **2002**. The methods to generate transgenic animals and transgene expression. *J. Biotechnol.* **98**, 145-160.

Houdbine LM. **2009**. Production of pharmaceutical proteins by transgenic animals. *Comp. Immunol. Microbiol. Infect Dis.* **32**, 107-121.

Huang B, Cui K, Li T, Wang XL, Lu F, Liu Q, da Silva FM and Shi D. **2008**. Generation of buffalo (*Bubalus bubalis*) transgenic chimeric and nuclear transfer embryos using embryonic germ-like cells expressing enhanced green fluorescent protein. *Reprod Domest Anim.* 1-6.

Kling J. **2009**. First US approval for transgenic animal drug. *Nature Biotechnol.* **27**. 302-304.

Kues WA, Niemann H. **2004**. The contribution of farm animals to human health. Review. *Trends Biotechnol.* **22**, 286-94.

Kues WA, Schwinzer R, Wirth D, Verhoeyen E, Lemme E, Herrmann D, Barg-Kues B, Hauser H, Wonigeit H and Niemann H. **2006**. Epigenetic silencing and tissue independent expression of a novel tetracycline inducible system in double-transgenic pigs. *FASEB J.* **20**, E1–E10.

Kuroiwa Y, Kasinathan P, Choi YJ, Naeem R, Tomizuka K, Sullivan EJ, Knott JG, Duteau A, Goldsby RA, Osborne BA, Ishida I, Robl JM. **2002.** Cloned transchromosomic calves producing human immunoglobulin. *Nature Biotechnol.* **20**, 889–894.

Lai L, Kolber-Simonds D, Park KW, Cheong HT, Greenstein JL, Im GS, Samuel M, Bonk A, Rieke A, Day BN, Murphy CN, Carter DB, Hawley RJ, Prather RS. **2002**. Production of alpha-1,3-galactosyltransferase knockout pigs by nuclear transfer cloning. *Science.* **295**, 1089–1092.

Laible G. **2009**. Enhancing livestock through genetic engineering- recent advances and future prospects. Review. *Comp. Immunol. Microbiol. Infect. Dis.* **32**. 123-137.

Lillico SG, Sherman A, McGrew MJ, Robertson CD, Smith J, Haslam C, Barnard P, Radcliffe PA, Mitrophanous KA, Elliot EA, Sang HM. **2007**. Oviduct-specific expression of two therapeutic proteins in transgenic hens. *Proc. Natl. Acad. Sci. USA.* **104**, 1771-1776.

Lo WH, Hwang SM, Chuang CK, Chen CY, Hu CY. **2009**. Development of hybrid baculoviral vector for sustained transgene expression. *Mol. Ther.* **17**, 658-666.

Lufino MM, Esder PA, Wade-Martins R. **2008**. Advances in high-capacity extrachromosomal vector technology: episomal maintenance, vector delivery, and transgene expression. *Mol. Ther.* **16**, 1525-1538.

Maga EA, Shoemaker CF, Rowe JD, Bon Durant RH, Anderson GB and Murray JD. **2006.** Production and processing of milk from transgenic goats expressing human lysozyme in the mammary gland. *J. Dairy Sci.* **89**, 518-524.

McCreath KJ, Howcroft J, Campbell KH, Colman A, Schnieke AE and Kind AJ. **2000.** Production of gene-targeted sheep by nuclear transfer from cultured somatic cells. *Nature.* **405**, 1066–1069.

Markaki M, Drabek D, Livadaras I, Craig RK. **2007**. Stable expression of human growth hormone over 50 generations of transgenic insect larvae. *Transgenic Res.* **16**, 99-107.

Müller M, Brenig B, Winnacker EL and Brem G. **1992**. Transgenic pigs carrying cDNA copies encoding the murine Mx1 protein which confers resistance to influenza virus infection. Gene. 121, 263-70.

Niemann H and Kues W. **2007**. Transgenic farm animals. *Reprod. Fertil. Dev.* **19**, 762-770.

Palmiter RD, Brinster RL, Hammer RE, Trumbauer ME, Rosenfeld MG, Birnberg NC, Evans RM. **1982**. Dramatic growth of mice that develop from eggs microinjected with metallothionein-growth hormone fusion genes Nature. 300, 611-615.

Phelps CJ, Koike C, Vaught TD, Boone J, Wells KD, Chen SH, Ball S, Specht SM, Polejaeva IA, Monahan JA, Jobst PM, Sharma SB, Lamborn AE, Garst AS, Moore M, Demetris AJ, Rudert WA, Bottino R, Bertera S, Trucco M, Starzl TE, Dai Y, Ayares DL. **2003**. Production of alpha 1,3-galactosyltransferase-deficient pigs. *Science.* **299**, 411-414.

Pursel VG, Pinkert CA, Miller KF, Volt DJ, Campbell RG, Palmiter RD, Brinster RL and Hammer RF. **1989**. Genetic engineering of livestock. *Science.* **244**, 1281-1288.

Redwan EM. **2009**. Animal derived pharmaceutical proteins. *J. Immun. Immunochem.* **30**, 262–290.

Ristevski S. **2005**. Making better transgenic models: conditional, temporal, and spatial approaches. Review. *Mol. Biotechnol.* **29**, 153-163.

Royer C, Jalabert A, Da Rocha M, Grenier AM, Mauchamp B, Couble P, Chavancy G. **2005**. Biosynthesis and cocoon-export of a recombinant globular protein in transgenic silkworms. *Transgenic Res.* **14**, 463-72.

Schnieke AE, Kind AJ, Ritchie WA, Mycock K, Scott AR, Ritchie M, Wilmut I, Colman A and Campbell KH. **1997**. Human factor IX transgenic sheep produced by transfer of nuclei from transfected fetal fibroblasts. *Science.* **278**, 2130–2133.

Simons JP, McClenaghan M, Clark AJ. **1987**. Alteration of the quality of the milk by expression of sheep beta-lactoglobulin in transgenic mice. *Nature.* **328**, 530-532.

Verma V, Gautam SK, Palta P, Manik RS, Singla SK, Chauhan MS. **2008**. Development of a pronuclear DNA microinjection technique for production of green fluorescent protein-expressing bubaline (*Bubalus bubalis*) embryos. *Theriogenology.* **69**, 655-665.

Wadhwa N, Kunj N, Tiwari S, Saraiya M and Majumdar SS. **2009**. Optimization of embryo culture conditions for increasing efficiency of cloning in buffalo (*Bubalus bubalis*) and generation of transgenic embryos via cloning. *Cloning Stem Cells.* **11**, 387-395.

Wang Z, Ruan J and Cui D. **2009**. Advances and prospect of nanotechnology in stem cells. *Nanoscale Res Lett.* **4**, 593-605.

Willadsen SM. **1986**. Nuclear transplantation in sheep embryos. *Nature.* **320**, 63–65.

Wilmut I, Schnieke AE, McWhir J, Kind AJ and Campbell KH. **1997**. Viable offspring derived from fetal and adult mammalian cells. *Nature.* 385, 810–813.

Wright G, Carver A, Cottom D, Reeves D, Scott A, Simons P, Wilmut I, Garner I, Colman A. **1991**. High level expression of active human alpha-1-antitrypsin in the milk. *Biotechnology.* **9**, 830-834.

Yang P, Wang J, Gong G, Sun X, Zhang R, Du Z, Liu Y, Li R, Ding F, Tang B, Dai Y, Li N. **2008.** Cattle mammary bioreactor generated by a novel procedure of transgenic cloning for large-scale production of functional human lectoferrin. *PLoS.* **3**, e3453

Zhang Y and Lu L. **2008.** Microinjection as a tool of mechanical delivery. *Curr. Opin. Microbiotechnol.* **19**, 506-510.

Zhu L, van de Lavoir MC, Albanese J, Beenhouwer DO, Cardarelli PM, Cuison S, Deng DF, Deshpande S, Diamond JH, Green L, Halk EL, Heyer BS, Kay RM, Kerchner A, Leighton PA, Mather CM, Morrison SL, Nikolov ZL, Passmore DB, Pradas-Monne A, Preston BT, Rangan VS, Shi M, Srinivasan M, White SG, Winters-Digiacinto P, Wong S, Zhou W and Etches RJ. **2005**. Production of human monoclonal antibody in eggs of chimeric chickens. *Nature Biotechnol.* **23**, 1159-1169.

—□□□—

10

Embryo Development Block, Epigenetic and Nuclear Reprogramming

"Successful reprogramming of a somatic genome to produce a healthy clone by SCNT is a rare event and the mechanisms involved in this process are poorly defined. When serial rounds of cloning are performed, blastocyst and full term development rates decline even further with the increasing rounds of cloning. Identifying the "cumulative errors" could reveal the basis of epigenetic reprogramming blocks in animal cloning. Genomics and microarray analysis of the donor nuclei have provided novel insights into the reprogramming of the genome, and have raised the hopes to improve nuclear transfer cloning for production of genetically engineered livestock species"

Introduction

As described in the preceding chapters, the animal cloning is becoming increasingly useful tool for a number of fundamental biological inquiries and for its role in pharmaceutical, medical, and agricultural fields. Despite decades of work the underlying reasons for regrettably low efficiency of the technique remain unresolved. This is highly probable that inefficiency is due to interaction of multiple factors and inefficient reprogramming of nuclear genome. Origin and quality of oocyte and donor nucleus and inept embryo culture conditions are other factors responsible for low developmental competence of the pre-implantation embryos. Attempts are being made to improve reprogramming of the donor nuclei and understanding the role of various genes involved in pluripotency during pre- and post-implantation development of the embryos.

What is embryo development block?

It is evident that mammalian development commences with the totipotent zygote which is capable of developing into all the specialized cells that make up the adult animal.

The preimplantation development of mammals is characterized by transition from oogenetic to embryonic genomic control, also called embryonic genome activation. As development proceeds after fertilization, maternally inherited mRNA molecules decay and embryogenesis becomes dependent on the expression of genetic information derived from the embryonic genome. The process of early embryonic development for bovine embryos begins as a "minor genome activation" at the 1-cell stage (Memilli and First, 2000) ascending to a "major genome activation" during the 8-cell to 16-cell stage (Whitworth *et al.*, 2004). In the absence of proper genome activation, the developing embryo will die because it can no longer support its essential develop-mental functions (Latham and Schultz, 2001; Han *et al.*, 2003). As development unfolds, cells of the early embryo proliferate and differentiate into early lineages, the pluripotent inner cell mass (ICM) and the tropthectoderm. The pluripotent cells from the inner cell mass of the blastocyst can be isolated, cultured and maintained in undifferentiated state as ES cells (Mitalipov and Wolf, 2009). ES and EG cells retain intrinsic activity to induce epigenetic reprogramming.

The embryo developmental block is a phenomenon not yet completely understood. It generally arises during the lag phase (the cell cycle which takes more time to be completed), and is possibly correlated with the genome activation (Meirelles *et al.*, 2004). Eight-cell embryo block is very common in mammalian embryonic development.

There are many factors that can interfere in embryo development in bovines causing embryo cleavage to cease, and most of the embryonic block occurs during the fourth or between the fourth and fifth cell cycle transition (Memili and First, 2000). This developmental block is observed in many species, with the peculiarity that it shows up at different stages (i.e. fourth cell cycle in cattle; fifth in rabbits; third to fourth in humans; second in mice, reviewed by Memili and First (2000) and in the fourth cell cycle in cats (Hoffert *et al.*, 1997).

Micro RNAs and maternally inherited developmental information seem to play an important role in controlling early cleavage of the mammalian embryo, thus decide the subsequent development. Two main mechanisms believed to cause the embryonic developmental block include inability to overcome the nuclear chromatin repression, and (i) activate transcription of important developmental regulatory genes and/ or (ii) to react to mechanical injuries caused by environment (Betts and King, 2001). With advances in genomics technology and development of powerful microarray systems to simultaneously elucidate the expression of a number of genes, novel strategies may be developed to resolve and increase the embryonic genomic activation, and thereby overcoming the developmental blocks in mammalian species.

Epigenetic reprogramming

Epigenetic regulation describes changes in chromatin. Epigenetic reprogramming is defined as any meiotic or mitotic alteration that does not result in change(s) in DNA sequence in genome, but will have significant impact on the development of an organism through altered gene-expression (Russo *et al.*, 1996). The most important form of epigenetic change in mammals is the symmetric methylation of cytosine in the 5′ position in CpG dinucleotides, or the deacetylation of certain lysine amino acids in histones. Both DNA methylation and histone deacetylation impair gene expression and are reversible (Reik and Walter, 2001). DNA methylation is thought to have two fold biological significance, gene regulation, transcriptional repression and 'genomic defense and structural fidelity' (Bestor, 2000; Robertson and Wolffe, 2000). The idea that histone acetylation-deacetylation plays a very important role in the process of reprogramming is further supported by the evidence that the SCNT success rate is greatly improved when trichostatin-A TSA is used. In addition, TSA treatment of reconstructed mouse embryos has resulted in the production of live offspring in a previously unclonable mouse strain (Kishigami *et al.*, 2007). In this manner, the epigenetic changes do not alter the nucleotide sequence of genes; nevertheless, have a great deal to do with whether or not a stretch of DNA is transcribed into RNA. Hence, in the process of development, it is different patterns of gene expression that make cell types functionally distinct from one another. Thus, epigenetic restrictions are an important component of individual cellular identity, the concepts articulated some time ago by Conrad Hal Waddington (1957) and in the absence of a modern mechanistic understanding of the underlying biochemistry.

For mammals, the studies show that epigenetic reprogramming is severely deficient in cloned embryos: most exhibit histone H3-K9 hypermethylation associated with DNA hypermethylation (Kang *et al.*, 2002; Santos *et al.*, 2003). As highlighted by the large offspring syndrome with *in vitro* conceived ovine and bovine animals, any disturbance during germ cell development or early embryogenesis has the potential to alter epigenetic reprogramming.

Epigenomics and epigenetics, therefore, provides a means of understanding how environmental factors might alter heritable changes in gene expression without changing DNA sequence and hence the origin of some diseases that are not explained by conventional genetic mechanisms. The epigenetic reprogramming of the donor cell nucleus is an important factor in the development of embryos and production of normal offspring derived by somatic cell nuclear transfer (*NT-SC*). In early embryos, this process makes it possible to erase gamete-specific methylation patterns and induce *de novo* methylation at defined developmental time-points. To clarify changes in DNA methylation in bovine *NT-SC* embryos, Sawai *et al.* (2009) examined satellite I sequences in bovine embryos derived *in vivo*, and by SCNT at the blastocyst (BC) and elongated (EL) stages. Because the EL stage embryo consists of the embryo disc (ED) and trophectoderm (TE), the methylation status of each part was analyzed with respect to the progress of differentiation. DNA methylation levels in *in vivo* embryos were increased during the elongation stage. In contrast, DNA methylation levels in *NT-SC* embryos remained unchanged in the ED and significantly decreased in the TE. Real-time PCR analysis showed that *Dnmt-1* expression in BC embryos derived by *NT-SC* was significantly lower than the *in vivo* produced embryos; thus, differences in the DNA methylation status may reflect transcript levels of *Dnmt-1*. The study suggested that the aberrant methylation level of bovine NT-SC embryos in the satellite I region is corrected as a result of demethylation and retention of methylation as the embryo develops and differentiates (Sawai *et al.*, 2009).

Crucial epigenetic reprogramming events occur during germ cell development and early embryogenesis. As highlighted by the large offspring syndrome with *in vitro* conceived ovine and bovine animals, any disturbance during germ cell development or early embryogenesis has the potential to alter epigenetic reprogramming. Therefore, the complete array of ART, starting from ovarian hormonal stimulation to uterine embryo transfer, could have a profound impact on the epigenetic state of *in vitro* produced individuals (Dupont *et al.*, 2009).

Reprogramming is essential in nuclear transfer

Reprogramming and nuclear transfer (NT) are the two terms frequently used when referring to animal cloning. Whereas the NT refers to the technique itself and is generally used to mean the introduction of a foreign nucleus into an (enucleated) matured oocyte, the reprogramming encompasses the complex mechanisms that follow NT, but is often used in an elusive manner (Jouneau and Renard, 2003). A recent and welcome definition describes reprogramming as "the molecular dominance of one distinct cell type over another, resulting in transformation of pliant nucleus to the dominant type" (Western and Surani, 2002). This definition is well adapted to the context of standard NT, and indicates the critical role of oocyte cytoplasm and underlines the importance of nuclear transformation for the developmental competence of the reconstructed oocytes.

There is a difference in the gene expression pattern between SCNT embryos and *in vitro*-fertilized embryos, suggesting that the low efficiency of SCNT is associated with incomplete reprogramming in the donor nuclei. Successful SCNT, therefore, requires that donor nucleus must be reprogrammed in order to become totipotent (Hiiragi and Solter, 2005).

The normal birth of a cloned animal precludes that a somatic cell chromatin introduced into a recipient oocyte has properly implemented remodeling, as well as epigenetic reprogramming during early embryogenesis. Till date the problem of epigenetic reprogramming remains the most important issue in cloning by nuclear transfer. For clones to complete development the genes normally expressed during embryogenesis but silent in donor cell must be reactivated so that the donor cells attain the stage of totipotency. Nuclear reprogramming mechanism is thought to involve two steps: (i) dedifferentiation of the differentiated donor cell to a totipotent embryonic state, followed by (ii) redifferentiation of cloned embryos to different somatic cell types during subsequent development (Yang *et al.*, 2007a, b). The first step of nuclear reprogramming refers to the erasure of the donor cell epigenetic pattern after NT and re-establishment of embryonic epigenetic characteristics and gene expression in the cloned embryo. The second step of the nuclear reprogramming refers to the redifferentiation of cloned embryos from a totipotent status to various differentiated states for tissue generation or organogenesis during post-implantation development (Yang *et al.*, 2007c). Reproductive cloning involves both the steps of reprogramming and is plagued by low efficiency; most cloned embryos die *in utero*, and the few that develop to term exhibit high incidences of abnormalities and short life span. Recent data from animal cloning suggest that inappropriate or aberrant expression of the gene is possibly the main cause of early embryo loss. It is also shown that aberrant genes expression patterns are associated with placental abnormalities in several species. On average, a laboratory loses 60 to 70% of *in vitro* matured oocytes due to the inability of the embryo to properly undergo cleavage and development to blastocyst stage (Lequarre *et al.*, 2003).

Data show that nuclei isolated from ES cells and embryonic blastomere have generated viable cloned animals with a significantly higher efficiency compared to those produced from adult somatic cells. This observation is consistent with the notion that genome of pluripotent stem cells is easier to be reprogrammed or requires less reprogramming than that of adult somatic cells. Also, fetal cells are more reprogrammable than adult somatic cells. Nevertheless, the successful *in vitro* developmental ability of SCNT embryo to the blastocyst stage does not assure the *in vivo* development after implantation. Hence, evaluation of reprogramming in donor nuclei appears to be most essential for predicting the *in vivo* developmental ability of SCNT embryos. For the practical purposes, the method should be non-invasive so that a SCNT embryo can be monitored for reprogramming and simultaneously saved for further development to term also.

Measuring and monitoring reprogramming

(Reprogramming efficiency i.e. the product of reprogramming ability and reprogrammability) can be detected and quantified at different levels of cellular information which are not currently well integrated (Oback, 2009). At genome level, the reprogramming has been measured as changes in DNA methylation (Kang *et al.*, 2003; Wang *et al.*, 2007; Yang *et. al.*, 2007a) and post-translational histone modifications (Santos *et al.*, 2003; Wang *et al.*, 2007; Yang *et al.*, 2007a, b). At the RNA level, epigenetic changes result in aberrant gene expression. Review of the literature shows that the spatiotemporal regulation of gene expressions is abnormal in cloned early cleavage stage embryos, blastocysts, and fetuses. This applies to both termination of the somatic and initiation of the embryonic gene expression program. Microarray comparison has shown that the donor-cell transcriptome is largely repressed by the late one-cell (Vassena *et al.*, 2007), and blastocyst stage (Smith *et al.*, 2005) which is an important indication of the differential reprogramming at various stages of embryo development. Microarray technology is useful in elucidating the expression of genes at various levels of embryonic development.

Improving nuclear reprogramming

Low developmental competence of bovine SCNT embryos is a global problem. Till date, little is known about the reprogramming, and reprogramming and NT are considered to be independent. Various strategies described below have been adopted, and are still under optimization to improve nuclear reprogramming thereby improving the NT.

Origin of oocytes and their reprogramming ability

The oocyte is the only cell in the adult body that can reprogram a terminally differentiated somatic cell back to a totipotent state (Cibelli *et al.*, 2002). This transformation requires a series of molecular events occurring correctly within a very short period of time, including a global resetting of the gene expression pattern of the donor nucleus. Based on extensive set of nuclear transfer experiments for comparing several different parameters in the NT to achieve optimal conditions in terms of the technique, Hiiragi and Solter, (2005) concluded that naturally ovulated oocytes were much better recipients of a nucleus compared to superovulated oocytes. Similarly, the abattoir-derived oocytes exhibit a low-developmental competence. This could be because in developing countries, primarily those animals are sold for slaughtering which are unable to breed and are dispensable to livestock owners. Irrespective of the origin of the oocytes, certain treatments of the IVM oocytes to derive cytoplasts have been found to increase the cloning efficiency. This indicates that NT efficiency can be increased to some extent by technical improvements.

Inducing nuclear reprogramming by spontaneous cell fusion

One well-known cell fusion phenomenon occurring *in vivo* is fertilization. The union of two pronuclei of an M-II oocyte and sperm creates the next generation with genetic diversity. Soon after fertilization, sperm chromatin is reprogrammed by nucleoplasmin because of exchange of sperm-specific basic proteins to histones (Philpott and Leno, 1992). The naturally fertilized eggs possess totipotential competence, as shown by their normal growth and differentiation into various types of cells, which form complex, functionally distinctive and multicellular tissues during the whole-body formation. Upon differentiation, the pluripotential competence is lost in the majority of somatic cells, though a subpopulation of somatic stem cells exists as multipotential cells, which maintain homeostasis in many tissues and organs by regeneration of replacement of cells (Tada, 2006).

The technique of cell fusion, which was pioneered by Henry Harris (1965) to generate heterokaryons between HeLa cells and rat lymphocytes, hen erythrocytes, and rabbit macrophages has proved to be a powerful procedure with numerous applications in cell biology, genetics and developmental biology and in the diverse fields of practical concerns such as medicine and agriculture. The spontaneous or induced cell fusion of two different types of cells (heterokaryons) results into intraspecific or interspecific hybrid cells, and provides a powerful tool for the molecular mechanisms of epigenetic reprogramming. In hybrid cells of somatic cells and pluripotential stem cells, including ES and embryonic germ cells, somatic nuclei acquire pluripotential competence.

The spontaneous fusion of the cells is a rare phenomenon and under *in vitro* culture conditions, has been found only in some cell lines and malignant cells. Induced cell fusion experiments have revealed the elevated molecular dominance and the reprogramming capabilities of embryonic cells compared to differentiated ones. Cell fusion can be induced by treatments with certain chemical agents such as calcium ions, lysolecithin, and polyethyleneglycol; by mediation by viruses such as paramyxoviruses, including Sendai virus (HVJ), oncornavirus, coronavirus, herpesvirus and poxvirus. The cells can also be fused by electrofusion. The electrofusion of the cells is important in inducing the totipotency in the terminally differentiated somatic cells in somatic cell nuclear transfer. Thus, nuclear reprogramming results from the resetting of somatic cell-specific epigenetic program by trans-acting factors present in unfertilized oocytes.

Nuclear reprogramming of somatic cells by cytoplast

The key role in the process of reprogramming has been attributed to the enucleated oocyte-cytoplast into which the somatic cell nucleus is transferred. Effective activation of a recipient oocyte and its compatibility with the nuclear donor are critical to the successful nuclear reprogramming during nuclear transfer. Derivation of viable cytoplasts through chemically-assisted enucleation is an accepted method for producing cloned embryos. Studies have shown that certain chemicals like demecolcine would greatly facilitate the removal of nuclear genetic materials from the IVM oocytes, thereby facilitating the efficient production of cytoplasts. The recent approaches of chemically-induced enucleation are based on the phenomenon that treatment of M-II phase with some cytoskeleton-relaxing agents induces an extrusion cone over the surface of the oocyte cytoplasm, which facilitates the removal of genetic material by mechanical aspiration or manual bisection of zona-free oocytes (Yang *et al.*, 2005).

Selection and treatment of donor nuclei

It has been found that embryonic and somatic nuclei require different recipient cytoplast environment for remodeling/ reprogramming, and this is likely due to the different cell cycle stage and profiles of molecular differentiation of the transferred donor nuclei. There is a renewed interest in using animal oocytes to reprogram human somatic cells. Researchers are using animal oocytes to reprogram human somatic cells, though the technique is controversial, and no animal-oocyte reprogrammed human cell lines are available. The systematic data clearly indicate that as embryos develop through pre-implamntation stages, their nuclei lose totipotency and that the efficiency of cloning depends on how far the donor cells have developed. This supports the notion of importance of reprogramming in nuclear transfer (Hiiragi and Solter, 2005). Studies using cow and rabbit oocyte cytoplasm clearly suggest that DNA methylation/ demethylation of the donor genome occur in a species-specific way, and that the ooplasm might lack the ability to demethylate repetitive sequences from other species (Chen *et al.*, 2006). Treatments of the donor nuclei, like, serum starvation, treatment with certain agents like trichostatin A, cytochalasin B, or roscovitine etc. have been found to improve their reprogramming, and subsequently the births of clones. In buffaloes also, demecolcine-assisted enucleation of IVM oocytes has been found suitable for deriving the cytoplasts, in SCNT as well as HMC embryos. Live clones have been produced from the transfer of these HMC embryos.

Culture medium and culture conditions of the embryos

In vitro culture environment plays an important role in somatic cell reprogramming. The embryos must reach a hatching stage of blastocysts suitable for transfer into recipient females. The process needs a suitable combination of culture media and culture conditions. Many strategies have been established to improve the efficiency of NT but relatively a few are focused on improving culture conditions. The effect of different culture media on pre-implantation development of mouse nuclear transfer embryos are investigated by various workers. Significance of a suitable culture medium in an *in vitro* embryo production system has been discussed in the preceding chapter on somatic cell nuclear transfer.

Applications of nuclear reprogramming

Cells of a multicellular organism are structurally and functionally heterogeneous owing to the differential expression of genes. Historically, the differential gene expression was thought to involve the genetic elimination of those genes that were silenced and the retention of others that were expressed in particular tissue (Hochedlinger and Jaenisch, 2006). The cloning experiments have unequivocally demonstrated that adult cells are genetically equivalent to early embryonic cells, and that differential gene expression is the result of reversible epigenetic changes that are gradually imposed on the genome during development and ageing of the cells.

Reprogramming through NT has been a unique tool for functionally testing the nuclear potency, and for distinguishing between genetic and epigenetic alterations of various donor cells (Hochedlinger *et al.*, 2004; Eggan *et al.*, 2004). Transformation of somatic nuclei to pluripotent nuclei via nuclear reprogramming has the potential to provide enormous benefits in the fields of agriculture, veterinary, basic sciences and clinical medicine by enabling the production of animal models for human diseases, bioreactors for production of human therapeutic proteins, organ donors for transplantation into human and livestock, which produce valuable food and neutraceuticals. Use of animals as live bioreactors in future may be helpful in cheaper production of therapeutically important biomolecules.

Another important contribution of reprogramming is the production of induced pluripotent stem (iPS) cells. These cells have potential applications in nuclear transfer cloning, improving reprogramming of somatic cells, *in vitro* production of gametes, *in vitro* meat production, and generation of tissues from iPS cells. Cell hybridization technology have the potential to make an important contribution to personal therapeutic applications without the need for cloning. Ultimately, manipulation of epigenotype may become a possibility through the use of reprogramming factors, once they have been identified. We then could envisage the production of clonal or tissue-specific stem cells from adult somatic cells, without the need for contribution from mammalian embryos (Tada, 2006). Such technology would have important implications for the production of donor cells for numerous clinical applications involving cell or tissue transplantation (Tada, 2006). A thorough understanding of these mechanisms will herald novel and exciting fields in the areas of both basic science and clinical medicine.

Critical concerns regarding reprogramming

Reports on generating iPS cells from non-embryonic tissues, but with characteristics identical to ES cells, has captured great attention in both scientific community as well as general public. Further, in the light of the continuing controversy over the use of ES and iPS cell, these reports have profound ramifications. A thorough investigation is needed on the topics, specifically, how convincing is the evidence supporting the claims made for the iPS cells? Should the iPS technique be considered, in the absence of any direct evidence for induction and reprogramming, as realistic alternative for SCNT to generate ES-like cells? A greater understanding of epigenetics, the process by which different patterns of gene expression are established, maintained and redirected, will continue to be enlightened by advances in cellular reprogramming. Recent studies have shown that reprogramming of microRNAs (miRNAs) occur in cloned bovine embryos. This could have profound implications for elucidating nuclear reprogramming in somatic cloning, as well as for the role of miRNAs in preimplantation mammalian development (Castro *et al.*, 2009).

Conclusion

Reproductive cloning involves both the steps of reprogramming and is plagued by low efficiency; most cloned embryos die *in utero*, and the few that develop to term exhibit high incidences of abnormalities and short life span. Further advances are expected to immensely improve genome reprogramming, thereby overcoming the developmental blocks in mammalian species during nuclear transfer cloning. Recent developments provide optimism that safe and viral-free iPS cells could be derived routinely in the near future. Next step will be to identify ways of assessing which iPS cell lines are sufficiently reprogrammed and safe for further applications.

References

Bestor TH. **2000**. The DNA methyl transferases of mammals. *Human Mol. Genet.* **9**, 2395-2402.

Betts DH and King WA. **2001**. Genetic regulation of embryo death and senescence. Review. *Theriogenology*. **55**, 171-191.

Castro FO, Sharbati S, Rodriguez-Alvarez LL, Cox JF, Hultschig C and Einspanier R. **2009**. MicroRNA expression profiling of elongated cloned and in vitro-fertilized bovine embryos. *Theriogenology*. (in press)

Chen T, Zhang YL, Jiang Y, Liu JH, Schatten H, Chen DY and Sun QY. **2006**. Interspecies nuclear transfer reveals that demethylation of specific repetitive sequences is determined by recipient ooplasm but not by donor intrinsic property in cloned embryos. *Mol. Reprod. Dev.* **73**, 313–317.

Cibelli J, Lanza RP, Campbell KHS and West MD (Eds). **2002**. Principles of Cloning (Academic Press, San Diego).

Dupont C, Armant DR, Brenner CA. **2009**. Epigenetics: definition, mechanisms and clinical perspective. *Semin. Reprod. Med.* **27**, 351-357.

Eggan K, Baldwin K, Tackett M, Osborne J, Gogos J, Chess A, Axel R and Jaenisch R. **2004**. Mice cloned from olfactory sensory neurons. *Nature*. **428**, 44-49.

Gurdon J. **1962**. The developmental capacity of nuclei taken from intestinal epithelium cells of feeding tadpoles. *J. Embryol. Exp. Morphol.* **10**, 622–640.

Han YM, Kang YK, Koo DB and Lee KK. **2003**. Nuclear reprogramming of cloned embryos produced in vitro. *Theriogenology*. **59**, 33-44.

Harris H. **1965**. Behaviour of differentiated nuclei in heterokaryons of animals cells from different species. *Nature*. **206**, 583-588.

Hiiragi T and Solter D. **2005**. Mechanisms of first cleavage specification in the mouse egg: is our body plan set at day 0? *Cell Cycle*. **4**, 661-665.

Hoffert KA, Anderson GB, Wildt DE and Roth TL. **1997**. Transition from maternal to embryonic control of development in IVM/ IVF domestic cat embryos. *Mol. Reprod. Dev.* **48**, 208-215.

Hochedlinger K and Jaenisch R. **2006**. Nuclear reprogramming and pluripotency. *Nature*. **441**, 1061-1067.

Hochedlinger K, Blelloch R, Brennan C, Yamada Y, Kim M, Chin L and Jaenisch R. **2004**. Reprogramming of melanoma genome by nuclear transplantation. *Genes Dev.* **18**, 1875-1885.

Jouneau A and Renard J. **2003**. Reprogramming in nuclear transfer. *Curr. Opin. Genet. Develop.* **13**, 486-491.

Kang YK, Park JS, Koo DB, Choi YH, Kim SU, Lee KK and Han YM. **2002**. Limited demethylation leaves mosaic-type methylation states in cloned bovine pre-implantation embryos. *EMBO J.* **21**, 1092-1100.

Kang YK, Lee KK and Han YM. **2003**. Reprogramming DNA methylation in pre-implantation stage: peeping with Dolly's eyes. *Curr. Opin. Cell Biol.* **15**, 290-295.

Kishigami S, Bui HT, Wakayama S, Tokunaga K, Van Thuan N, Hikichi T, Mizutani E, Ohta H, Suetsugu R, Sata T and Wakayama T. **2007**. Successful mouse cloning of an outbred strain by trichostatin A treatment after somatic cell nuclear transfer. *J. Reprod. Dev.* **53**, 165–70.

Latham KE and Schultz RM. **2001**. Embryonic genome activation. *Front. Biosci.* **6**, D748-7D59.

Lequarre AS, Marchandise J, Moreau B, Massip A and Donnay I. **2003.** Cell cycle duration at the time of maternal zygotic transition for in vitro produced bovine embryos: effect of oxygen tension and transcription inhibition. *Biol. Reprod.* **69**, 1707-1713.

Meirelles MV, Caetano AR, Watanabe YF, Ripamonte P, Carambula SF, Merighe GK and Garcia SM. **2004**. Genome activation and developmental block in bovine embryos. *Anim. Reprod. Sci.* **82-83**, 13-20.

Memilli E and First NL. **2000**. Zygotic and embryonic gene expression in cow: a review of timing and mechanisms of early gene expression as compared with other species. *Zygote.* **8**, 87-96.

Mitalipov S and Wolf D. **2009**. Totipotency, pluripotency and nuclear reprogramming. *Adv. Biochem. Eng. Biotechnol.* **114**, 185-199.

Oback B. **2009**. Cloning from stem cells: different lineages, different species, same story. *Reprod. Fertil. Dev.* **21**, 83-94.

Philpott A and Leno GH. **1992**. Nucleoplasmin remodels sperm chromatin in *Xenopus* egg extracts. *Cell.* **69**, 759-767.

Reik W and Walter J. **2001**. Genomic imprinting: parental influence on the genome. *Nat. Rev. Genet.* **2**, 21–32.

Robertson KD and Wolffe AP. **2000**. DNA methylation in health and disease. *Nature Rev. Genetics.* **1**, 11-19.

Russo VEA, Martienssen RA and Riggs AD. **1996**. Epigenetic Mechanisms of Gene Regulation. Cold Spring Harbor, NY: Cold Spring Harbor Laboratory Press.

Santos F, Zakhartchenko V, Stojkovic M, Peters A, Jenuwein T, Wolf E, Reik W and Dean W. **2003**. Epigenetic marking correlates with developmental potential in cloned bovine preimplantation embryos. *Curr. Biol.* **13**, 1116-1121.

Sawai K, Takahashi M, Moriyasu S, Hirayama H, Minamihashi A, Hashizume T and Onoe S. **2009**. Changes in the DNA methylation status of bovine embryos from the blastocyst to elongated stage derived from somatic cell nuclear transfer. *Cloning Stem Cells.* (in press)

Smith SL, Everts RE, Tian XC, Du F, Sung LY, Rodriguez-Zas SL, Jeong BS, Renard JP, Lewin HA and Yang X. **2005**. Global gene expression profiles reveal significant nuclear reprogramming by the blastocyst stage after cloning. *Proc. Natl. Acad. Sci. USA*. **102,** 17582-17587.

Tada T. **2006**. Nuclear reprogramming. An Overview. *Methods Mol. Biol.* **348,** 229-236.

Vassena R, Han Z, Gao S, Baldwin DA, Schultz RM and Latham KE. **2007a.** Tough beginning: alterations in the transcriptome of cloned embryos during the first two cycles. *Dev. Biol.* **304,** 75-79.

Waddington CH. **1957**. The Strategy of the Genes: A Discussion of Some Aspects of Theoretical Biology. London: Allen and Unwin.

Wang F, Kou Z, Zhang Y and Guo S. **2007**. Dynamic reprogramming of histone acetylation and methylation in the first cell cycle of cloned mouse embryos. *Biol. Reprod.* **77,** 1007-1016.

Western PS and Surani MA. **2002**. Nuclear reprogramming- alchemy or analysis. *Nature Biotechnol.* **20,** 445-446.

Whitworth K, Springer GK, Forrester LJ, Spollen WG, Ries J, Lamberson WR, Bivens N, Murphy CN, Mathialagan N, Green JA and Prather RS. **2004**. Developmental expression of 2489 gene clusters during pig embryogenesis: an expressed sequence tag project. *Biol. Reprod.* **71,** 1230-1243. Erratum in: Biol Reprod. 2004. 71(6), 2092. Mathialigan, Nagappan [corrected to Mathialagan, Nagappan].

Yang F, Hao R, Kessler B, Brem G, Wolf E and Zakhartchenko V. **2007**a. Rabbit somatic cell cloning: effects of donor cell type, histone acetylation status and chimeric embryo complementation. *Reproduction*. **133,** 219-230.

Yang J, Yang S, Beaujean N, Niu Y, He X, Xie Y, Tang X, Wang L, Zhou Q and Ji W. **2007**b. Epigenetic marks in cloned rhesus monkey embryos: comparison with counterparts produced in vitro. *Biol. Reprod.* **76,** 36-42.

Yang X, Smith SL, Tian XC, Lewin HA, Renard JP and Wakayama T. **2007**c. Nuclear reprogramming of cloned embryos and its implications for therapeutic cloning. *Nature Genet.* **39,** 295-302.

Yang XY, Zhao JG, Li HW, Li H, Liu HF, Huang SZ and Zeng YT. **2005**. Improved efficiency of cloning by fusion of somatic cells to hybrid oocyte recipients. *Theriogenology*. **64,** 1263-1272.

—□□□—

11

Genomics in Livestock Production

"Farm animal genomics is of interest to a wide range of researchers because of utility derived from understanding how genomics and proteomics function in organisms. Applications of these methods for livestock research are rapidly becoming as extensive as that for research in model organisms and humans. Research in genomics and proteomics technologies may help animal genetic selection more effectively. While the advances in the field promise to offer key insights into molecular biology, much of the available information remains underutilized because of the lack of scalable approaches for integrating signals across large and diverse data sets"

Introduction

In commercial and scientific livestock production systems two types of breeding practices are relevant. First, 'classical or conventional' genetics is concerned with statistical correlations between phenotype and genotype- the genes themselves are not identified, but patterns of inheritance are understood within a probabilistic framework. Second, a molecular approach to genetics aims to identify the specific genetic elements that are biochemically involved in producing a particular phenotype – a particular (typically small) sequence of DNA within the comparatively vast genome of the organism is recognized as a gene contributing to a trait (Gibbs *et al.*, 2009). Both classical and molecular genetics can be used to guide breeding decisions, although only a molecular understanding of genetics can enable the practice of genetic modification. Increasingly, the molecular genetics research is using a genomic approach, in which a great many (e.g. thousands) of genetic elements are analyzed for their subtle, networked contributions to traits of interest.

Livestock breeding industries in developed countries have achieved major genetic progress, boosting their efficiency and competitiveness for improved sustainability. Presently, many of these technologies are being used in research on searching and identifying the candidate genes and disease diagnosis in economically important animals. Integrating the information encoded in the genome to applied nutrition and breeding and ultimately, augmenting the livestock production is the goal. However, recent consumer and societal demands for better and safe food quality, as well as animal welfare and environmental concerns, have explicitly raised a demand for identification of new livestock breeding traits and reproduction management strategies. These new challenges exceed the capacities of traditional livestock selection and crossbreeding tools, because of low heritabilities, measurement difficulties and higher cost. Modern genomics tools can possibly offer solution to most of these problems.

Clearly, the potential genetic benefits in livestock production and health from the developments in molecular biology are very great. Livestock genomics has the strength of both human and model-species genetics because researchers can exploit both the abundant genetic variation between divergent breeds and the variation that is segregating within breeds. Collectively, the *'omics'* applications in the fields of biology and medicine are becoming a reality, beyond the conventional genomic sequencing. Applications of genomics are likely to extend to almost all the areas related to livestock production and management. This chapter highlights some of the recent hot areas for genomics interventions that have immediate impact on livestock health and production.

Genomics for exploiting rumen microbiome

In the past 2 decades, research in both academia and industry has made use of molecular biology, genetics, and comparative genomics, which has led to the development of key technologies for the discovery of novel biomolecules for health and industrial applications. Genome-scale efforts have led to the identification of numerous molecular targets. A wide variety of approaches are being used in the search for novel biocatalysts, antibiotics and genes from the novel microbes. There is already a substantial record of classical genetics and biotechnological contributions to livestock nutrition. These include single cell protein production, the genetic modification of nutrient value of forages, and the provision of a range of probiotic and antibiotic feed additives.

Studying mammalian (human and animals) gut ecosystem is an emerging and important area of genomics. This is because the gut ecosystem is one of the most complex and systematically coordinated microbial ecosystems studied so far. The review of literature reveals that gut ecosystem may offer an inexhaustible source of microbial enzymes, biotherapeutics, genes and genetic engineering tools and novel products for applications in animal health industrial development (Selinger *et al.*, 1996; Singh *et al.*, 2001; Flint *et al.*, 2008; Morrison *et al.*, 2009). The herbivores, particularly the ruminants and termites retain within their gastrointestinal tract a microbiome that specializes in the rapid hydrolysis and fermentation of lignocellulosisc plant biomass (Morrison *et al.*, 2009). The symbiosis of this extended genome plays a pivotal role in host homeostasis, derivation of nutrients and energy from the crude dietary resources. Collectively, these symbionts are responsible for the digestion of high roughage diets, detoxification of a number of plant secondary metabolites (PSMs) and synthesis of volatile fatty acids (VFAs), vitamins and microbial proteins which are ultimately utilized by the host. Due to the presence of unique obligate anaerobes (fungi, protozoa, bacteria and archaea) and continuous formation of microbial products, the rumen has been regarded as fountain head of valuable hydrolytic enzymes (hemicellulases, xylanases, cellulases, endoglucanases, acetyl xylan esterases and tannase etc.) that has increasing demand in feed (plant biomass saccharification, ensuring the supply of critical nutrients to the host), textile and pulp and paper processing (Selinger *et al.*, 1996; Palackal *et al.*, 2007; Flint *et al.*, 2008; Singh *et al.*, 2009).

Metagenomics is enabling microbial ecologists and physiologists to characterize genomes of the microbes which are not readily isolated or able to grow *in vitro*, leading to the discovery of thousands of novel microbial species and products. Metagenomic studies include mammalian and environmental samples, where microbes play important roles in health, agriculture, and the environment. The human gut, for example, is predicted to contain around 100 trillion bacteria, with colonization of the gastrointestinal tract beginning at birth. Similarly, gut microbes play an essential role in ruminants as the later are dependent on anaerobic microbial species to degrade cellulose and other dietary ingredients. The microbial species in the rumen interact in a complex manner resulting in a remarkable metabolic capacity. Understanding the composition and function of this community will be of paramount importance. Culture-independent genomics and metagenomics methods, therefore, provide unique insights into this complex gut ecosystem, and using the same for augmenting livestock production **(Table-1)**. Rumen of the buffaloes adapted to digest high fiber forages can harbor potentially elite microbes for use as direct fed microbials (probiotics) and microbial enzymes (prebiotics) to improve nutrient utilization in other livestock species like poultry and swine.

Genomics and modification of livestock feed resources

This is another important and integrated aspect of improving livestock productivity using gene-based technologies. The additional livestock products demand will be fulfilled only if the nutrient requirements of meat and milk producing animals are fulfilled.

The most significant application of molecular genetics in agriculture to date have been the development of genetically modified (GM) crops, which are aimed at reducing production cost. Additional applications of GM technology include improved synthesis of amino acids in plants, lowered lignifications of fibrous forages, lower mycotoxin contamination in grains through reduction in plant borer damage, and reduced environmental pollution through better utilization of dietary forage nitrogen and phosphorus. Increased synthesis of PSMs in the plant could be utilized for augmenting livestock rumen and health production (Singh *et al.*, 2003).

Applying genetics to improve fertility

The significance of genetic variation in fertility improvement is globally accepted. Heritability of fertility traits commonly used in animal breeding is relatively low, as consequence of the large unexplainable residual variation in statistical models trying to predict traits like calving intervals and pregnancy rate at the individual animal level (Veerkamp and Beerda, 2007). There is overwhelming evidence that increasing genetic merit for yield without considering genetic merit for fertility may actually reduce fertility (Veerkamp *et al.*, 2003). Gene expression profiling has emerged as powerful tool for gaining knowledge on which genes and their products determine specific traits, and it is anticipated that this technology will have an important role in improving fertility (Dawson, 2006; Wolf *et al.*, 2006). Developments in veterinary genomic medicine will likely explain much of what is now considered idiopathic male infertility. This is because our understanding of the genetic defects that cause infertility is no longer confined to chromosomal aneuploidies and single-gene defects (e.g. cystic fibrosis and congenital absence of the vas deferens).

New techniques for assessing genomic influences on phenotypic traits, such as those related to dairy cow fertility are becoming available at seemingly ever increasing speed. High-throughput technologies will generate vast amounts of quickly accessible information on static aspects of the genome, including SNPs, QTL, genes and the polymorphism in them (Smits *et al.*, 2008), as well as on dynamic aspects such as DNA cytosine methylation patterns, histone modification marks and gene expression profiles. Undoubtedly, this will greatly increase our knowledge on genomic influences on dairy cow fertility, though effective handling, integration and interpretation of the data may prove challenging. For the near future, a contribution of high-throughput technologies to breeding strategies is expected as in part they have been adapted already, but conventional breeding strategies based on increasingly sophisticated phenotypic traits are likely to predominate for some years (Veerkamp and Beerda, 2007).

Genomics- assisted livestock breeding

The integrated research area of modern animal genomics is now moving from its sequencing era into an integrative functional genomics era. Application of molecular biological techniques has potential to revolutionize the animal health and production. One of the 'grand challenges' in modern biology is to understand the genetic basis of animal phenotypic diversity within and among species. This is attributed to thousands of years of selective breeding of domestic animals which has created a diversity of phenotypes among breeds that is only matched with those observed among the species in nature. Domestic animals, therefore, constitute a unique resource for understanding the genetic basis of phenotypic variation. When the genome sequences of domestic animals become available, the identification of the mutations that underlie the transformation from a wild to a domestic species will become a realistic and important target (Andersson and Georges, 2004).

Genomics-based animal selection predicts breeding values for a large number of haplotypes across the entire genomes that are derived from (combinations of) marker alleles. Compared to traditional breeding practice, genomic-based information yields a considerable increase in selection responses for juvenile animals for which phenotypic records are not yet available (Meuwissen *et al.*, 2001), and can also reduce the costs of a breeding program up to 90% (Schaeffer, 2006). Companies have begun to develop genotyping tests to identify those animals with the greatest potential for tenderness, fat content and weight gain. In 2006, two Dutch breeding companies announced the implementation of genomic selection in dairy cattle (HG) using 3000 SNPs and in broilers (Nutreco) using 20,000 SNPs. Several other breeding companies are expected to shortly initiate animal selection and implementation programs based on genetic makeup of the individual animals. It is argued that these tests provide a more rigorous system to grade meat quality than the subjective grading of meat by inspectors after slaughter. The importance of genetic selection for fertility are likely to increase further (Veerkamp and Beerda, 2007).

Genome-wide panels of SNPs have recently been used in domestic animal species to map and identify genes for many traits and to select the genetically desirable livestock. This has led to the discovery of the causal genes and mutations for several single-gene traits but not for complex traits (Goddard and Hayes, 2009). Next-generation high-throughput DNA sequencing techniques are opening fascinating opportunities in the life sciences. At the moment genomes of important animal species are already sequenced or being sequenced. Unfortunately, the blueprint of structural organization of billions of bases of genomic DNA does not really tell what all the nucleotides sequences or genes do, how the different genes interact to perform specific cellular activities, what is going wrong in disease states, or how an organism ages. This is where the functional genomics comes into play. The overall goal is not simply to provide a catalogue of all the genes and the information about their functions, but to elucidate how the genome components work together to make the cells and organisms functioning. In the functional genomics, the real-time gene expression can be analyzed at the RNA (transcriptomics) or protein (proteomics) levels or, in those cases in which expression is affected by alterations in the copy number of the genes, hence, also at the level of genome.

The reproduction related gene expression profiling studies cover a broad range of reproductive processes at the level of the pituitary, ovaries, oviducts, uterus and embryo and underscore the importance of steroid biosynthesis, oxygen radical metabolism, regulation of cell cycles structure and morphology (including apoptosis), inter and intracellular communication, tissue remodeling and (immune-related) protein secretion

Additionally, the functional genomics tools like nutrigenomics, metabolomics and metabonomics are assisting appreciably in understanding the gene regulation in response to dietary and environmental factors. The research has focused on the mechanisms controlling gene-expression and impact of biological, gut microbial and external factors on gene-expression (e.g. genetic determinants and nutritional factors, respectively) in certain tissues involved in metabolism, reproduction, growth and production (milk and meat) traits of the animals (Cassar-Malek *et al.*, 2008). The ability to predict total genetic merit of animal using molecular markers may offer opportunities to completely redesign the livestock breeding and management programs (Sellner *et al.*, 2007). High-throughput arrays and comparative genomics are being used to detect numerous transcripts simultaneously, and compare their expression in a single experimental analysis (Bonnet *et al.*, 2008).

The fast growing sequence information of animal genomics provides opportunities to improve animal health traits by genomics-assisted breeding (Smits *et al.*, 2008). In this concern, the search for quantitative trait loci (QTL) in large ruminants has been extensively pursued with emphasis on growth and milk production related traits. While genetic markers for QTL can be used to choose animals for selective breeding programs, the most effective markers are those which have functional mutations within the desirable trait genes. In modern genomic era numerous DNA sequences have been annotated in economically important farm animal species, and the data generated may have applications in evaluation and analysis of similar gene sequences in other livestock species.

Genomics applications in bubaline production and reproduction

The functional genomics studies have been originally performed in humans, primates, laboratory species, marine species (aquaculture) and some livestock species. The information derived is being applied to domestic animal species. Genomics studies in beef and dairy cattle have focused on oocyte maturation, non-regressed and regressed corpus luteum tissue, oviduct epithelial cell function, the endometrium during the estrous cycle, pre-implantation embryonic development, and embryo-induced transcriptome changes of the endometrium in the pre- and periattachment period. The knowledge acquired may offer enormous potential in improving and exploring the current reproductive technologies among buffaloes.

Understanding of spatio-temporal profiling of the regulatory genes, elucidation of the molecular basis of transcriptional control during the pre-implantation development is of utmost importance for diagnosing, intervening and eliminating or reducing the abnormalities associated with cloned or *in vitro* produced embryos. Molecular developmental dynamics, *viz.*, aberrant epigenetic regulation, transcriptional repression of key pluripotency genes in cloned buffalo embryos may provide useful insight into the basic causes of embryonic developmental arrest and low efficiency of SCNT in the species, and enable necessary interventions for improving the existing protocols.

A thorough understanding of the cellular and molecular mechanisms regulating the early embryonic developments is important to improve efficiency of *in vitro* embryo production in bubaline species. Also, identification of molecular markers of oocyte eligibility, and improving the oocyte developmental-competence are the fundamental issues in boosting buffalo assisted reproduction. With the advent of functional genomics era, it is now possible to analyze the transcriptome of a cell, and the similar efforts should focus on identification of novel predictors and regulators of oocyte-developmental competence in buffaloes. Using the BlueChip bovine cDNA microarray platform and quantitative real-time PCR analysis, Kandil *et al.*, (2009) have reported that about 104 transcripts were differentially expressed between immature and *in vitro* matured buffalo oocytes. Transcription factors (ZFP91), M-phase mitotic cell cycle (MPHOSPH9), growth factor (BMP15), cDNA binding (HMGN2) were found to be upregulated in immature oocytes, whereas IVM oocytes had elevated expressions of genes involved in cytoskeleton (ACTB), hydrogen ion transporting (ATP6V1C2) and structural constituent of ribosome (RPS27A) (Kandil *et al.*, 2009). Further, for understanding the chromatin remodeling activities in cloned buffalo embryos, Suteevun *et al.*, (2006) examined the expression profiles of key regulatory genes involved in DNA and histone modifications in single metaphase oocytes, nuclear transfer- and IVF-derived embryos. Quantitative analysis of the genes revealed that cloned buffalo embryos exhibited significantly higher mRNA transcripts of embryonic chromatin remodeling genes (*DNMT1, DNMT3A, DNMT3B, HAT1* and *HDAC1*), which are shown to repress gene transcription during early embryonic stages (Suteevan *et al.*, (2006). These studies on identification of candidate genes in buffalo oocytes and embryos are novel, and may explain the genetic basis of poor developmental competence of the cloned embryos.

Challenges and opportunities

Domestic animals constitute a unique resource for understanding the genetic basis of phenotypic variation. Gene profiling studies are relatively novel and the possibilities to perform gene expression profiling studies are growing rapidly. Applications of molecular markers and statistical analysis of the data on molecular variations is widening our capacity to precisely characterize the genetic variations of livestock breeds (Toro *et al.*, 2009). With the completion of buffalo genome sequencing project, the molecular markers may offer important applications in future, to correct, complete or implement buffalo pedigree recording. In the immediate future, structural genomics will be restricted to development of better genome maps, including linkage maps, radiation hybrid maps, contig maps and transcript maps. Once the information is at our disposition, the next challenge will be to identify single genes with QTL regions and dissect their contribution to the traits of economic interest. It is now possible to estimate the genetic merit of animals by genomic selection, which uses genome-wide SNP panels as markers and statistical methods that capture the effects of large numbers of SNPs simultaneously. This approach is expected to double the rate of genetic improvement per year in many livestock systems (Goddard and Hayes, 2009).

Further, applications of genomics discoveries that are emerging from the application of next generation DNA sequencing technologies may have a striking impact on animal genetics and genomics. These new methods have expanded previously focused readouts from a variety of DNA preparation protocols to a genome-wide scale and have fine-tuned their resolution to single base precision. The sequencing of RNA also has transitioned and now includes full-length cDNA analyses, serial analysis of gene expression (SAGE)-based methods, and noncoding RNA discovery. The genomics and proteomic technologies combined with the bioinformatics tools that support the interpretation of gene functioning and protein expression facilitate an exciting starting point for the development and validation of new livestock management strategies. Taken together, an astounding potential exists for these technologies to bring enormous change in genetic and biological research and to enhance our fundamental biological knowledge as tools for the improvement of bubaline reproductive performance.

Conclusion

The use of genetic technology in animal production has moved from a theoretical concept to the beginnings of practical application. Genomics is setting new paradigms in research approaches within biological sciences, and will be a major force in enhancing the rate of progress in understanding biological systems and exploiting them for development products. The rapid developments that occurred in the field of livestock genome analysis over the past several years require the cooperation of specialists from diverse fields. Scientists with expertise in molecular biology, veterinary medicine, biochemistry, and quantitative genetics need to collaborate to solve the problems in buffalo breeding and husbandry. Rumen molecular techniques have great potential for enhancing rumen efficiency through introduction of improved fermentation activities thereby improving the overall health and immunity of the animals. Genomics and molecular biological tools should be used as methods of choice to gain insights into biology of complex traits, and overcoming the key problems currently associated with advanced ARTs. The field of buffalo assisted reproduction will benefit efficiently from the close collaborations between buffalo breeders, molecular biologists, nutritionists and interdisciplinary scientific communities that share research and technology expertise.

Table 1. A summary of salient applications of genomics to augment livestock production through nutritional interventions

Targets	Future prospects
Novel hydrolytic enzymes	Identification and characterization of enzymes for use in animals feeds, pulp and paper, and textile industry
	Applications as prebiotics for enhanced utility of plant biomass for improving rumen productivity and applications in poultry and swine (Singh *et al.*, 2008a,b)
Novel microbial species	Establishing new *in vitro* culture conditions for therapeutically or nutritionally relevant novel gut flora for use as probiotics or direct-fed microbials (DFMs) for animals
	Developing the strategies for biomonitoring of the transinoculated gut-based DFMs
	Studying the gut ecosystem of the animals exhibiting natural adaptation to the diets containing astringent PSMs, identification of novel gut microbes and microbial enzymes and using them for overcoming dietary PSMs toxicity
	Studying the interactions among different microbial consortia and between host and the microbes

Novel genes and genetic engineering tools	Exploiting the gut microbes as resources of novel genes, restriction enzymes, and plasmids as tools for genetic engineering of the resident normal flora
Therapeutics	Identification of the rumen bacteriocins for use as rumen modulators or for suppression of GI pathogens, applications in processed dairy foods and meat products
Methanogenesis	Identification of the novel methanogens in the GI tract, manipulation of the rumen for lowering green house gas emissions
	Inducing animals immune system-mediated antibodies against rumen methanogens, developing vaccines against the selected methanogens

Table 2. Applications of molecular biological and genomics tools to improve bubaline assisted reproduction and health

Techniques	Proposed prospects
Genetic improvement and management	Genotyping of indigenous buffalo breeds adapted to prevailing low-input and extreme climatic conditions
	Prediction of total genetic merit of the selected genotypes, and re-designing the buffalo breeding management- programs based on whole-genome selection (WGSL)
	Understanding the role of epigenetic mechanisms influencing the economically important phenotypes
Follicliculogenomics & embryogenomics	Gaining insights into molecular and cellular differentiation and oocyte development during folliculogenesis
	Analysis of transcriptome during maturation of oocyte for exploring the genes involved in the process and the specific checkpoints regulating acquisition of full competence for supporting fertilization and further embryonic development (Assou *et al.*, 2006)
	Gene-expression profiling in the oocyte and early embryos for optimization of IVF and IVEP (Kandil *et al.*, 2009; Aswal *et al.*, 2008)

	Identification of oocyte-specific genes which are functionally indispensable for the developmental competence of oocyte
	Revisiting oocyte-somatic cell interactions for identifying the predictors and regulators of occyte developmental competence
	Exploring the molecular basis of differential developmental competence of *in vitro vs. in vivo* matured oocytes (Kandil *et al.*, 2009)
	Identification of key regulatory and pluripotency genes associated with the developmental competence and viability of the cloned embryos
Stem cell technology	Markers-assisted differentiation of trophoblast and/ or the endoderm cells overgrowing the ICM-derived ES cells
	Understanding the molecular basis, and expression pattern of stemness genes in buffalo ES cells, and elucidating the mechanisms involved in retaining undifferentiated state of buffalo ES cells as well as mechanisms involved in their differentiation
Animal cloning	Establishing criteria for selection of donor nuclei based on their genetic suitability for improved nuclear transfer cloning
	Elucidating epigenetic patterning of the cloned embryos, as indicators of their viability to possibly resolve the embryonic developmental blocks in the species
	Developing strategies for the improved expression of key pluripotency genes through innovations *in vitro* embryo culture media and systems
Health and management	Exploring nutrigenomics and metabolomics tools for nutrient-guided expression of the genes of interest in the target tissues
	Exploring the molecular basis of thermoregulation in the animals performing under extreme tropical conditions, and establishing and disseminating the buffalo herds tolerant to thermal stress

	Identifying the important genes and genetic pathways involved in thermoregulation in the elite breeds exhibiting a natural adaptation to extreme environmental conditions
	Elucidation of the novel biomolecules, genes and the genetic pathways involved in the immune responses to mastitis
	Generating the baisc knowledge of the biology of reproduction and fertilty, that will result in the design of more successful breeding programs for improving the sustainability of bubaline production and animal welfare

References

Andersson L and Georges M. **2004**. Domestic-animal genomics: deciphering the genetics of complex traits. *Nat. Rev. Genet.* **5**, 202-212.

Assou S, Anahory T, Pantesco V, Le Carrouer T, Pellestor F, Klein B, Reyftmann L, Dechaud H, De Vos J, Hamamah S. **2006**. The human cumulus-oocyte complex gene-expression profile. *Hum. Reprod.* **21**, 1705-1719.

Aswal AP, Raghav S, De S, Thakur M, Goswami SL, Datta TK. **2008.** Expression stability of two housekeeping genes (18S rRNA and G3PDH) during in vitro maturation of follicular oocytes in buffalo (*Bubalus bubalis*). *Anim. Reprod. Sci.* **103**, 164-171.

Bonnet A, Dalbies-Tran R, Sirard MA. **2008.** Opportunities and challanges in applying genomics to the study of oogenesis and folliculogenesis in farm animals. *Reproduction.* **135**,119-128.

Cassar-Malek I, Picard B, Bernard C, Hocquette J. **2008**. Application of gene expression studies in livestock production systems: a European perspective. *Aust. J. Exp. Agric.* **48**, 701-710.

Dawson KA. **2006**. Nutigenomics: feeding the genes for improved fertility. *Anim. Reprod. Sci.* **96**, 312-322.

Flint HJ, Bayer EA, Rincon MT, Lamed R and White BA. **2008**. Polysaccharide utilization by gut bacteria: potential for new insights from genomic analysis. *Nature Rev. Microbiol.* **6**, 121-131.

Gibbs D, Holloway L, Gilna B and Morris C. 2009. Genetic techniques for livestock breeding: restructuring institutional relationships in agriculture. *Geoforum.* (in press)

Goddard ME and Hayes BJ. **2009**. Mapping genes for complex traits in domestic animals and their use in breeding programmes. *Nature Rev. Genet.* **10**, 381-391.

Kandil OM, Ghanem N, Abdoon A, Holker M, Phatsara C, Schellander K, Tesfaye D. **2008.** Transcriptional analysis of buffalo (*Bubalus bubalis*) oocytes during in vitro maturation using bovine cDNA microarray. *Reprod. Domest. Anim.* (in press).

Meuwissen TH, Hayes BJ and Goddard ME. **2001**. Prediction of total genetic value using genome-wide dense marker maps. *Genetics.* **157**, 1819–1829.

Morrison M, Pope PB, Denman SE and McSweeney CS. **2009**. Plant biomass degradation by gut microbiomes: more of the same or something new? *Curr. Opin. Biotechnol.* **20**, 358-363.

Palackal N, Lyon CS, Zaidi S, Luginbuhl P, Dupree P and Goubet F. **2007**. A multifunctional hybrid glycosyl hydrolase discovered in an uncultured microbial consortium from ruminant gut. *Appl. Micobiol. Biotechnol.* **74**, 113–124.

Schaeffer LR. **2006**. Strategy for applying genome-wide selection in dairy cattle. *J. Anim. Breed. Genet.* **123**, 218–23.

Selinger LB, Forsberg CW and Cheng KJ. **1996**. The rumen: a unique source of enzymes for enhancing livestock production. *Anaerobe.* **2**, 263–284.

Singh B, Bhat TK and Singh B. **2001**. Exploiting gastrointestinal microbes for livestock and industrial development. *Asian-Aust. J. Anim. Sci.* **14**, 567–586.

Singh B, Bhat TK and Singh B. **2003**. Potential therapeutic applications of some anti-nutritional plant secondary metabolites. *J. Agirc. Food Chem.* **51**, 5579-5597.

Singh B, Gautam SK and Mukesh M. **2009**. Exploring the rumen ecosystem for improved livestock productivity: a metagenomic overview. *Indian Dairyman.* **61** (9), 50-55.

Singh B, Gautam SK, Verma V, Kumar M and Singh, B. **2008a**. Metagenomics in animals gastrointestinal tract- potential biotechnological prospects. *Anaerobe.* **14**, 138-144.

Singh B, Bhat TK, Sharma OP and & Kurade NP. **2008b**. Metagenomics in animal gastrointestinal tract- a microbiological and biotechnological perspective. *Indian J. Microbiol.* **48**, 216-227.

Sellner EM, Kim JW, McClure MC, Taylor KH, Schnabel RD and Taylor JF. **2007.** Board invited review: applications of genomic information in livestock. *J. Anim. Sci.* **85**, 1-11.

Smits MA, Rebel JMJ, te Pas, MFW, Hulst MM. **2008.** Exploiting genomics to improve animal health. *Dev. Biol. (Basel).* **132**, 27-33.

Suteevun T, Smith SL, Muenthaisong S, Yang X, Parnpai R and Tian XC. **2006**. Anomalous mRNA levels of chromatin remodeling genes in swamp buffalo (*Bubalus bubalis*) cloned embryos. *Theriogenology.* **65**, 1704-1715.

Toro MA, Fernandez J and Caballero A. **2009**. Molecular characterization of breeds and its use in conservation. *Livestock Sci.* **120**, 174-195.

Veerkamp RF and Beerda B. 2007. Genetics and genomics to improve fertility in high producing dairy cow. Theriogenology. **68S**, S266-S273.

Veerkamp RF, Beerda B and van der Lende T. **2003**. Effects of genetic selection for milk yield on energy balance, levels of hormones, and metabolites in lactating cattle and possible links to reduced fertility. *Livest. Prod. Sci.* **83**, 257-275.

Wolf E, Arnold GJ, Bauersachs S, Blum H, Frohlich T, Hiendleder S. **2006**. Functional genome research-new strategies to improve fertility in cattle. *Zuchtungskunde*. **78**, 428–39.

—□□□—

Glossary of the Terms Used

Acrosome: A protrusion on the anterior end of a sperm cell containing hydrolytic enzymes (e. g. hyaluronidase) that enable the sperm to penetrate oocyte.

Adaptation: In the evolutionary sense, some heritable feature of an individual's phenotypes that improve its probability of survival and reproduction in the existing environment.

Amnion: The innermost membranous sac enclosing the embryo of an amniote; it becomes filled with amniotic fluid. One of the four amniote extra embryonic membranes; derived from somatopleure (combination of ectoderm and somatic mesoderm). Used mainly for the suspension of the embryo.

Amniotic fluid: The fluid surrounding the developing fetus that is found within the amniotic sac contained in the mother's womb.

Archea: One of the three domains of life. Archaea are unicellular microorganisms and have a combination of characteristics unlike of the other two domains.

Artificial insemination: An assisted reproductive technology in which semen with live sperm is collected from the males, then diluted and processed and used in fresh or frozen form to breed females at proper time with the help of instruments.

Bacterial artificial chromosome (BAC): An artificial high capacity cloning vector used to clone DNA fragments of 100 to 300 kb in bacterial or eukaryotic cells.

Bacteriocin: A bacterial ribosomally synthesized protein or peptide molecule exhibiting a narrow range of antagonism, generally against the closely related bacterial species. Bacteriocins confer the host additional tool to compete with other microorganisms in a specific niche.

Bioinformatics: A fast developing branch of science comprising of information technology, computer science, and biological data. Bioinformatics concerns the development of new tools for analysis of enormous genomic and molecular biological data including sequence analysis, phylogenetic inferences, genome database organizing and mining, optical computation and holographic memory, pattern recognition and image analysis, biologically inspired computational models, genetic algorithms, machine learning, modeling of complex biological and molecular processes and artificial life etc.

Biological diversity or biodiversity: It is that part of nature which includes the differences in genes among the individuals of a species, the variety and richness of all the plant and animal species at different scales in space, locally, in a region, in the country and the world, and various types of ecosystems, both terrestrial and aquatic, within a defined area.

Biomarker: The official NIH definition of a biomarker is: "a characteristic that is objectively measured and evaluated as an indicator of normal biologic process, pathogenic process or pharmacologic responses to a therapeutic intervention".

Biomass: Refers to total dry weight of all organisms in particular sample, population, or area.

Biomes: Recognizable community units formed by the interaction of regional climate, regional biota, and substrate. The same biome units generally can be found on different continents at the same latitudes that produce about the same weather conditions and where topography is similar. Biomes are the largest land community recognized.

Biopharming: It is the method of developing and exploiting the genetically modified organisms (microorganisms, plant or animals) for producing recombinant proteins or other biomolecules for therapeutic applications.

Blastocoel: Fluid-filled cavity found in the interior of a blastocyst.

Blastocyst: The blastocyst is the developing stage of an embryo which has the form of a hollow ball with a solid sphere, the inner cell mass cells, positioned at one end of the inner aspect of the hollow ball. The blastocyst is composed of only three defined cell types, namely the trophectoderm, primitive endoderm and inner cell mass (ICM) cells.

Blastomere: One of the cells formed by mitotic division of the fertilized egg making up the embryo or blastula.

Buffer: A solution system capable of resisting changes in pH even when acid or base is added, consisting of conjugate acid-base pair in which ratio of proton acceptor to proton donor is near unity.

Candidate or susceptibility genes: These are the genes with functionally important variations, having heightened sensitivity to diet. They may be chronically activated during a disease state and have been previously demonstrated to be sensitive to dietary intervention, and have important implications in hierarchical role in biological cascades. The candidate genes are believed to be associated with polymorphisms that are highly prevalent in populations.

Capacitation : Capacitation is process involving multiple signaling transduction pathways in which a sperm undergoes a series of maturational changes in the female reproductive tract. Capacitation can be induced *in vitro*.

Cell culture: A growth of cells *in vitro*; although the cells cultured *in vitro* are capable of proliferating but they do not organize into tissue.

Chimera: An organism comprising of more than one typeof somatic cells.

Chromatin: The DNA and its associated histone proteins around which it coils in the nucleus.

Chromosome: A linear end-to-end arrangement of genes and other DNA sometimes associated with protein and RNA. Chromosome carries genetic material in cells and viruses. Exists as a circle of DNA in prokaryotes, a DNA or RNA molecule in viruses, and a linear nucleoprotein complex in eukaryotes

Clone: An organisms that is the carbon copy of donor parent, and is produced through asexual reproduction. A clone may have more than one mother, but have no contribution from the father.

Cloning: It is the art of constructing or producing a cloned molecule or organism. The term cloning is used in different contexts (e.g., gene cloning, therapeutic cloning, adult animal cloning, etc.) in biological science.

Cloning vector: Generally, a recombinant DNA molecule originating from a virus, plasmid or the eukaryotic cells, into which another fragment of DNA can be integrated without loss of capacity for self-replication when introduced into the host cell. Examples of vectors include plasmids, cosmids, phage vectors, bacterial artificial chromosome, yeast artificial chromosome etc.

Cryopreservation: Method of preserving the biological material under extreme low temperature. There are two major techniques for cryopreservation: slow-freezing and vitrification. The major difference between two is total avoidance of ice formation in vitrification.

Cryoprotective agent: A cryoprotective agent (CPA) is a substance that is commonly added to freezing medium to prevent the biological material from freezing injuries i.e. damage due to ice formation. In most cases, little or no survival of cells is obtained without the application of CPA. The CPA may be permeating or non-permeating types.

Cytoplast: The oocyte without nuclear genetic material is called cytoplast.

Differentiation: The changes in a cell shape and physiology associated with the production of the final cell types of a particular organ or tissue.

Ectopic expression: The occurrence of gene expression in a tissue in which it is normally expressed. Ectopic expression can be induced by the juxtaposition of novel enhancer elements to a gene during genetic manipulation of transgenic organisms.

Ejaculation: The process of sudden release of fluid, especially of semen from the body; the semen is an ejaculate.

Electrofusion: Fusion of cells under the influence of electric field; generally induced by the electric field induced polarity in cell membranes.

Embryo: An entity or organism or fertilized egg in early stages of development before hatching from its shell. The term is also used to describe the early stages of fetal growth, from conception to the eighth week of pregnancy.

Embryo developmental block: It is a phenomenon not yet completely understood, in which the embryo suddenly stops developing *in vitro* at certain stage. It generally arises during the lag phase and is possibly correlated with the genome activation.

Embryo transfer technology: Embryo transfer technology consists of a series of integrated sequential steps from oocyte donor selection, donor treatment for superovulation, insemination of the superovulated donor, embryo recovery, preimplantation embryo development and evaluation, recipient selection and synchronization, embryo transfer, through to healthcare and management of surrogate mothers carrying *in vitro* produced, cloned or transgenic fetuses.

Enucleation: It is the process of removal of nuclear genetic materials from an *in vitro* matured oocyte to produce cytoplast. Enucleation is important to maintain the ploidy in offspring resulting from nuclear transfer of the somatic cells.

Epigenetics: The process, by which different patterns of gene expression are established, maintained and redirected. Epigenetics provides a means of understanding how environmental factors might alter heritable changes in gene expression without changing DNA sequence and hence the origin of some diseases that are not explained by conventional genetic mechanisms

Epigenomics: seeks to define the influence of changes to gene expression that are independent of gene sequence and encompasses multiple mechanisms by which DNA transcription is altered in various tissues and at different times without changing the underlying gene sequence.

Estrus synchronization: The process of scheduling a group of probable embryo-recipients to come into estrus at approximately the same time as embryo-donor animals which are being superovulated with exogenous hormones. Estrous synchronization also gives the dairy and beef producers the opportunity to capture the economic benefits of AI in groups for better management at calving.

Extinction: In molecular biology it is the phenomenon of transcriptional downregulation of the genes in cells. Also refers to disappearance of a living species from ecosystem.

FACS: The fluorescent activated cell sorter equipment used to separate the cells into different fractions based on different fluorescent stain taken by the varied cell types/ organelles.

FDA: Food and Drug Administration, the U.S. agency responsible for regulation of biotechnology food products e.g GM crops. The major laws under which the agency has regulatory powers include the Food, Drug, and Cosmetic Act; and the Public Health Service Act.

Fertility: Fertility in dairy animals is defined as the ability of the animal to conceive and maintain pregnancy if inseminated at the appropriate time relative to ovulation

Fetus: The unborn young from the eighth week of pregnancy to birth as in human beings and bovines; an organism in the stage of development that follows the embryonic stages.

FISH: Fluorescent *in situ* hybridization, a physical mapping approach that uses fluorescein tags to detect hybridization of probes with metaphase chromosomes and with the less-condensed somatic chromatin.

Functional genomics: Mainly deals with elucidation and characterization of overall pattern of the gene expression of a cell, tissue or an organism.

Gastrointestinal: Refers to the entire alimentary canal including oral cavity, stomach and intestine.

Genetic resources, animal: Animal genetic resources are those animal species that are used, or may be used, for the production of food and agriculture, and the populations within each of them.

Genetically engineered organisms: Any plant or animal life whose genes are directly altered through incorporation of exogenous genes from a completely different type of organism rather than through cross breeding or cross-pollination of closely related organisms as in hybrids.

Genetics: Refers to the study of inheritance and variation.

Genome: The sum of all genetic information in an organism. The entire genetic complement of a virus, prokaryote, mitochondrion or chloroplast, or the haploid nuclear genetic complement of a eukaryotic species.

Genomic imprinting: Refers to an epigenetic mark that distinguishes parental alleles and results in a monoallelic, parental-specific expression pattern in mammals

Genomics: Genomics is the scientific study of structure, function and interrelationship of both individuals and the genome. The major objective of genomics is to gain an exhaustive understanding of the structure and function of the genomes, and is classified into two broad areas i.e. structural genomics and functional genomics.

Genotype: The specific allelic composition of a cell, either of the entire cell or more commonly for a certain gene or a set of genes that confer overall merit to an organism.

Germplasm: The sum total of all hereditary material in a single interbreeding species.

Haplotype: A haplotype, a contraction of the phrase "haploid genotype", is asset of closely linked genetic markers present on one chromosome which tend to be inherited together. Another way to think about it is that a haplotype is half of a genotype. Some haplotypes may be in linkage disequilibrium. A haplotype can be identified by patterns of SNPs (see later), which can be used to identify complex genetic variations or inheritable diseases.

Histone acetylation: The process of cellular enzymatic (histone acetyltransferases) addition of acetyl groups to the histone, thus facilitating gene transcription.

HMC: Handmade cloning: The technique of producing nuclear transfer cloned embryos. The techniques excludes the use of micro tools and micromanipulators for preparation of cytoplasts and transfer of donor nuclei to the cytoplast.

Host: An organism that contains or supports the life of another organism; or a cell or an organism that is used as recipient of exogenous genetic material to express the proteins encoded by exogenously introduced genes.

Immunosurgery: A method used for isolation of cells from inner cell mass cells using the principles of immune reactions.

Implantation: The end of free-floating period of an embryo with its attachment to the lining of womb so that it makes a direct contact with maternal body for the exchange of nutrients and metabolites.

In vitro: An experimental procedure performed outside the body.

In vivo: An experimental procedure performed inside the body.

Induced pluripotent stem (iPS) cells: The pluripotent cells derived from any differentiated cell type through ectopic expression of transcription factors. Originally, through retroviral expression of *Oct4, Sox2, Klf4* and c-*Myc*; other combinations and ways of generating iPS cells are now available.

Insertional mutation: Disruption in the activity of a gene due to insertion or cloning of exogenous gene in it.

IVF: *In vitro* fertilization, an important assisted reproduction tool which involves aspirating, maturing and fertilizing the oocytes outside the female genital tract (*in vitro*), generally in laboratory, for production of healthy and viable embryos for transfer to disseminate the superior genotype.

Metabolome: The collection of the complete complement of all low molecular weight (<1500 Daltons, Da) metabolites found in cell, tissue or organism, under a given set of conditions.

Metabolomics: Quantitative measurement of dynamic multiparametric metabolic response of living organism or system to pathophysiologic stimuli or genetic modification. While proteomics tends to focus on big molecules, the metabolomics focuses on small molecules.

Metabonomics: The global profiling of metabolites in different living systems. It is quantitative measurement of the dynamic multiparametric metabolic responses of living systems to pathpophysiological stimuli or genetic modifications.

Metagenomics: The genomics-based study of genetic material recovered directly from environmentally-derived samples without laboratory culture. Metagenomics is an emerging field focused on characterizing the structures, functions and dynamic operations of microbial communities sampled in their native habitats.

Micro RNAs (miRNAs): These are short, non-coding RNAs that negatively regulate target mRNAs via binding to their 3' untranslated regions (UTRs).

Microarray: Sets of miniaturized chemical reaction areas that may also be used to test DNA fragments, antibodies, or protein(s), by using a chip having immobilized target and hybridizing them with probed sample. The color obtained from hybridization is then scanned and data is analyzed by software to find expression levels or to provide inferences

Microbial ecology: The study of the interactions between living microorganisms and their environment.

Microbiome: A term to represent the genomes of microbial community in combination. For instance, gut microbiome refers to the gut microbial ecosystem.

Micromanipulation: The art of embryo cloning which involves dissection, suction and insertion of nuclear material or intracellular organelles usually in early embryo-developing stages.

Multipotent cells: The cells capable of differentiating into multiple cell types but within a certain lineage e.g. hematopoietic stem cells (HSCs).

Nuclear transfer: See somatic cell nuclear transfer

Nutriceutical: Common food product that have been modified (potentially by genetic engineering) to have enhanced nutritional value and/ or pharmaceutical applications.

Nutraceuticals: In the United States, it is defined as any substance that is a food or part of food and provides medical or health benefit; such products can range from a nutrient, dietary supplement, specialized diets to the genetically engineered foods. In Canada, it is defined as a product isolated or purified from foods, sold in the forms not usually associated with food, and demonstrated to have a physiological benefit against chronic disease.

Nutrigenetics: The nutrigenetics aims to understandhow genetic makeup of an individual coordinates the response to diet.

Nutrigenomics: It is the study of the effects of foods and food constituents on gene expression. It is about how our DNA is transcribed into mRNA and then to proteins and provides the basis for understanding biological activity of food components.

Nutrigenomics: The branch of functional genomics dealing with the impact of nutrients and environmental factors on expression of the genes.

Nutritional genomics: The study of how dietary factors affect total function of the genome including differences in response based on individual genetic makeup and interaction of dietary factors with genome and its subsequent proteomic and metabolic changes.

Oocyte: The female gamete.

Oogenesis: The process of formation of an oocyte.

Ovum pick-up: An assisted reproduction method of recovery of oocyte from the living female under the guidance of ultrasound scanning or laparoscopy. While ultrasound is usually transvaginal, laparoscopy is abdominal.

Passaging: Method of subculturing the animal cells

Permeabilization: Denotes formation of pores in the plasma membrane or partial complete removal of plasma membrane.

PGD: Pre-implantation genetic diagnosis is an established procedure of embryo genetic diagnosis, and is used to analyze embryos genetically before their transfer into the uterus.

Placenta: The membranous vascular fetal organ that develops in female mammals during pregnancy, lining the uterine wall and partially enveloping the fetus, to which it is attached by the umbilical cord.

Pluripotency: The potential of a cell to differentiate and develop into more than one type of mature cells depending on environment but can not give rise to new body. The adult somatic cell can be made pluripotent through exogenous introduction of certain factors, and the process is called as induced pluripotency.

Pluripotent cells: The cells capable of differentiating all germ layers (endoderm, mesoderm and ectoderm) and the germline but not extra embryonic tissues (e.g., inner cell mass, ES, EG, EpiSC, iPS cells).

Primary culture: A cell or tissue culture started from the material taken directly from an organism, as opposed to that from an explant from an organism.

Primordial germ cells: The cells residing in the early genital ridge are another source of cell lines with similar properties as stem cells.

Probe: A substance, such as DNA or RNA, that is radioactively labeled or otherwise marked and used to detect or identify another substance in a sample.

Proteomics: Proteomics deals with the study of production and functioning of proteins in biological system. It occupies a central platform among the '*omics*' tools that have more recently been adapted and applied to nutrition and health research in order to deliver biomarkers for health and comfort as well as to discover beneficial food bioactives.

Quantitative trait locus: A gene affecting the phenotypic variation in continuously varying traits like skin color, body weight, milk production etc.

Quantitative trait: A measurable trait that shows continuous variation; a trait that cannot be classified into a few discrete classes.

Reprogramming: The process that enables or forces the cell into a state of gene expression very similar to zygotic nucleus or embryonic stem cells.

Restriction endonuclease: Endonuclease that recognizes a particular short DNA sequence which they cleave. They help to protect the host cells from viral infection and are used in work with DNA. The enzyme makes two incisions, one through each of the phosphate backbone of the double helix without damaging the constituent bases.

SCNT: Pl. see somatic cell nuclear transfer.

Semen: Semen or male germplasm is the complete-discharge of the male genital tract occurring at the time of ejaculation. It is a white, opaque, creamy fluid, occasionally yellowish green due to the pigment carotene, and consists of cellular part i.e. spermatozoa or sperms and the fluid parts, known as seminal plasma.

Seminal plasma: The liquid portion of semen i.e. seminal plasma is nothing but the secretions of accessory sex glands such as the prostrate, seminal vesicles and cowper's glands. Seminal plasma presents an ideal medium for the viability and mobility of the sperm.

Serum: The liquid component of clotted blood. Serum lacks the formed elements and the clotting factors (which plasma includes), but retains the electrolytes and soluble biomolecules including antibodies.

Sex determination: The genetic or environmental process by which sex of an individual is established.

Sex-preselection: The ability to decide or determine the sex of embryo prior to implantation in uterus.

Sexual reproduction: The process where two gametes, sperm and oocytes, fuse to form one hybrid fertilized totipotent cell, the zygote.

SNP: Single nucleotide polymorphism

Somatic cell fusion: The fusion of somatic cells (usually from different organisms) using exogenous factors such as viruses, chemical agents (glycol) or electric impulse (electrofusion). The technique is used in somatic cell genetics, or formation of a hybrid cell (heterokaryon) or formation of cell-cytoplast couplet as in nuclear transfer cloning

Somatic cell nuclear transfer: A laboratory procedure in which a cell/ cell's nucleus is removed and placed into an enucleated oocytes (cytoplast), so that the genetic information in the somatic cell nucleus controls the resulting cells. The process is carried out for reprogramming of the donor nucleus, so that it may behave like a zygote and develop into an embryo

Somatic cell: A cell that is not destined to become a gamete under natural conditions; a cell whose genes cannot be passed on to future generations.

Spermatogenesis: The process of sperm production.

SRY: Sex determining region Y, the sex switch or testis-determining factor in mammals located on the Y-chromosome.

Stem cells: An unspecialized cell that may give rise to differentiated cells. Stem cells are characterized by their ability to proliferate indefinitely. Embryonic stem cells are derived from fertilized embryos which have undergone a limited number of cell division and differentiation from zygote.

Structural genomics: Deals with structural organization of the genome. Broadly, the structural geneomics deals with DNA sequencing, unraveling DNA complexity, DNA variability and polymorphisms, and genotyping.

Superovulation: Superovulation is a term used to describe the drug-induced production of multiple eggs for use during assisted reproduction. The process is used to produce multiple embryos.

Synchronization, estrus: Synchronization of estrus refers to the process of scheduling a group of embryo-recipient females to come into estrus at approximately the same time as embryo-donor animals which are being superovulated with exogenous hormones.

Tissue culture: Maintenance or growth of tissue, organ primordial, or the whole or part of an organ in vitro so as to preserve its architecture and function.

Totipotency: The ability of a cell to proceed through all stages of development, thus to form a normal fetus and offspring.

Totipotent cells: The cells capable of developing into a complete organism or differentiating into any of its cells or tissues (e.g. zygote and early blastomeres).

Transcriptomics: The branch of genomics dealing with formation of RNA from DNA.

Transgenics: The term originally refers to the stable introduction of exogenous genetic material into their germ line. The process can lead to gain of function (knock-in) or loss of function (knock-out) and spatial-temporal gene transfer (conditional knock-out) in an animal.

Trophectoderm: The outer most and surrounding cell layer in the blastocyst, which gives rise to placenta after implantation.

Unipotent cells: The cells that have the capacity to differentiate into only one type of cell (e.g. spermatogonial stem cells).

Vitrification: The method of cryopreservation thatis based on procedures that minimize the formation of intra-cellular ice crystals when an embryo and the surrounding solution are "glassified."

Yeast artificial chromosome (YAC): A high efficiency cloning vector.

Zygote: The unique diploid cell formed by the fusion of two haploid cells (often egg and a sperm) that will then divide mitotically to create a differentiated diploid organism.

—□□□—

Index

D

E

F

S

T

—□□□—

ABOUT THE AUTHORS

Dr. Prem Singh Yadav graduated from college of Agriculture HAU, Hisar, Haryana in 1985. He obtained his M.Sc. in 1987 and Ph.D. in 1991 from Department of Animal Production Physiology, College of Animal Sciences, CCSHAU, Hisar. Dr Yadav joined Agricultural Research Services of ICAR in 1993 as Scientist, Animal Physiology. He got selected as Senior Scientist in 2000 and promoted as Principal Scientist in 2008 at Central Institute for Research on Buffaloes, Hisar (Haryana). His research work remained mainly on embryo biotechnology and livestock stem cell research. As Overseas Associate of DBT, Government of India, he studied potential of bovine embryonic stem cells in Institute of Animal Sciences, Germany. He is working on adult stem cells in buffaloes with grants from DBT. Dr. Yadav has organized ICAR sponsored short course and summer schools. He has published 17 papers in international and national journals, edited compendia of conferences and trainings and guided M.Vsc and Ph.D. students.

Dr. Birbal Singh is Senior Scientist at Indian Veterinary Research Institute Regional Station, Palampur, HP. He did his M.Sc. and Ph.D. (Animal Biotechnology) from National Dairy Research Institute, Karnal, Haryana. He joined Indian Council of Agriculture Research (ICAR) as Scientist in 1996. His research expertise includes biotechnology of mammalian gut ecosystem and animal reproductive biotechnology (somatic cell nuclear transfer, stem cell biology and cryobanking). He has also worked on molecular characterization and toxicological investigations in fishes, and has published 20 research papers, reviews, invited book chapters, and technical articles in peer reviewed high-quality national and international journals.

Dr. Inderjeet Singh graduated from the College of Veterinary Sciences, Haryana Agricultural University, Hisar in December 1982. His M.Vsc thesis research laid the foundation for embryo biotechnology work in the university. He was awarded the coveted Commonwealth fellowship to pursue Ph.D from the University of Liverpool, U.K., which he completed in 1994. Dr. Singh started his professional teaching and research career from Haryana Agricultural University in 1985 and continued till his selection as Principal Scientist at the Central Institute for Research on Buffaloes, Hisar in 2001. Dr. Singh was elected Fellow of the National Academy of Veterinary Sciences and was selected for Dr. C.R. Sane Oration by the Indian Society for Study of Animal Reproduction (ISSAR) in 2004. He also served as the General Secretary ISSAR from 2005 to 2008. He has guided several M.Vsc and Ph.D. students. He has served on several important committees of the Government of India, and selection committees for teaching, research, extension and administrative positions. He has published two books in Hindi, two practical manuals besides editing several proceedings and compendia of scientific conferences, trainings etc. He served as Editor of the Haryana Veterinarian and on the Editorial Board of Intas Polivet. In pursuit of scientific agenda, Dr. Singh has visited over a dozen foreign countries.

Dr. R.K. Sethi graduated in Animal Sciences in 1974 and completed Masters Degree in the field of Animal Genetics & Breeding from Haryana Agricultural University; Hisar during 1976. He joined ICAR in Agricultural Research Services in 1977 as Scientist S-1 at the National Dairy Research Institute, Karnal. He acquired Ph.D. degree in Animal Genetics & Breeding in 1983 as in service candidate from NDRI Karnal. Selected as Scientist S-3 (Principal Scientist) in 1988 and joined at Central Institute for Research on Buffaloes, Hisar (Haryana). He has been looking after the charge of Project Coordinator (Buffalo) of the Network Project on Buffalo Improvement with 15 centers located in different parts of the country since 1993. He is Director of Central Institute for Research on Buffaloes, Hisar since Nov., 2007. He has 69 research papers at his credit in national and international journals and more than 100 presentations in Conferences/Symposia/Seminars etc. He has guided 5 M.Sc. and 2 Ph.D. students. He has visited several countries to participate in trainings/ seminars etc. including Bulgaria, USA, Vietnam, Egypt, Brazil and China.